EAI/Springer Innovations in Communication and Computing

Series Editor

Imrich Chlamtac, European Alliance for Innovation, Ghent, Belgium

The impact of information technologies is creating a new world yet not fully understood. The extent and speed of economic, life style and social changes already perceived in everyday life is hard to estimate without understanding the technological driving forces behind it. This series presents contributed volumes featuring the latest research and development in the various information engineering technologies that play a key role in this process. The range of topics, focusing primarily on communications and computing engineering include, but are not limited to, wireless networks; mobile communication; design and learning; gaming; interaction; e-health and pervasive healthcare; energy management; smart grids; internet of things; cognitive radio networks; computation; cloud computing; ubiquitous connectivity, and in mode general smart living, smart cities, Internet of Things and more. The series publishes a combination of expanded papers selected from hosted and sponsored European Alliance for Innovation (EAI) conferences that present cutting edge, global research as well as provide new perspectives on traditional related engineering fields. This content, complemented with open calls for contribution of book titles and individual chapters, together maintain Springer's and EAI's high standards of academic excellence. The audience for the books consists of researchers, industry professionals, advanced level students as well as practitioners in related fields of activity include information and communication specialists, security experts, economists, urban planners, doctors, and in general representatives in all those walks of life affected ad contributing to the information revolution.

Indexing: This series is indexed in Scopus, Ei Compendex, and zbMATH.

About EAI - EAI is a grassroots member organization initiated through cooperation between businesses, public, private and government organizations to address the global challenges of Europe's future competitiveness and link the European Research community with its counterparts around the globe. EAI reaches out to hundreds of thousands of individual subscribers on all continents and collaborates with an institutional member base including Fortune 500 companies, government organizations, and educational institutions, provide a free research and innovation platform. Through its open free membership model EAI promotes a new research and innovation culture based on collaboration, connectivity and recognition of excellence by community.

Xiaolin Jia • Thurasamy Ramayah • Hui Zhang
Editors

2nd International Conference on Big Data, IoT, and Cloud Computing

ICBICC 2024

Springer

EAI
RESEARCH MEETS INNOVATION

Editors
Xiaolin Jia
Southwest University of Science and Technology
Sichuan, China

Hui Zhang
Southwest University of Science and Technology
Sichuan, China

Thurasamy Ramayah
School of Management
Universiti Sains Malaysia
Penang, Malaysia

2nd International Conference on Big Data, IoT, and Cloud Computing (ICBICC) 2024

ISSN 2522-8595 ISSN 2522-8609 (electronic)
EAI/Springer Innovations in Communication and Computing
ISBN 978-3-031-93569-5 ISBN 978-3-031-93570-1 (eBook)
https://doi.org/10.1007/978-3-031-93570-1

This Springer imprint is published by the registered company Springer Nature Switzerland AG
The registered company address is: Gewerbestrasse 11, 6330 Cham, Switzerland

Preface

The 2024 2nd International Conference on Big Data, IoT, and Cloud Computing (ICBICC 2024) was held successfully in Xishuangbanna, China, from December 30, 2024 to January 1, 2025.

It aims to provide a global platform for scientists, scholars, and professionals from areas within Big Data, IoT, and Cloud Computing to exchange ideas, share experiences, and present the latest advancements in their respective domains.

This year, we received 62 submissions, which underwent a rigorous double-blind peer-review process. Ultimately, 24 high-quality papers were accepted for publication, a significant achievement that would not have been possible without the dedicated efforts of the proceedings editors and the reviewers. We extend our heartfelt gratitude to them for their invaluable contributions.

We sincerely appreciate all participants for their valuable contributions. Special thanks are extended to the speakers and committee members for their steadfast support. Their dedication and commitment have played a vital role in making this conference a success. We are also grateful for the unwavering support from individuals and institutions which has greatly contributed to the conference.

Our deepest gratitude goes to all who have contributed to ICBICC 2024. Their dedication and expertise are invaluable. Hopefully, ICBICC will eventually serve as a valuable resource for researchers and scholars in the fields of Big Data, IoT, and Cloud Computing.

Sichuan, China — Xiaolin Jia
Penang, Malaysia — Thurasamy Ramayah
Sichuan, China — Hui Zhang

Committee List

Conference Chairs

Xiaolin Jia, Southwest University of Science and Technology, China
Hui Zhang, Southwest University of Science and Technology, China

Program Chairs

Ramayah Thurasamy, Universiti Sains Malaysia, Malaysia
Massimo Marchiori, University of Padua, Italy
Qingming Zhang, Southwest University of Science and Technology, China
Changyong Xie, Southwest University of Science and Technology, China
Hua Fan, University of Electronic Science and Technology of China, China

Program Co-Chairs

Yihong Chen, China West Normal University, China
Gang He, Southwest University of Science and Technology, China
Haocun Wu, South China University of Technology, China

Publication Chair

Xiaofang Huang, Southwest University of Science and Technology, China

Finance Chair

Jangala. Sasi Kiran, Lords Institute Of Engineering and Technology, India

Publicity Chair

Xujian Zhao, Southwest University of Science and Technology, China

Award Chair

Yufang Chen, Southwest University of Science and Technology, China

Program Committee

Gyu Myoung Lee, Liverpool John Moores University, UK
Hong Huang, Sichuan University of Science & Engineering, China

Xiaolin Zhu, China West Normal University, China
Xiaoming Gao, Southwest University of Science and Technology, China
Xuan Guo, University of North Texas, USA
Yusheng Liu, Huawei Technologies Co., Ltd, China

Contents

Blockchain Architecture and System Design for Resource-Limited Satellite-Terrestrial Integrated Network

Shufan Jia, Jianjun Chen, and Hailong Kang

1 Introduction

In recent years, satellite network has been developing rapidly, and many satellites are about to be launched into space. Satellite network has advantages such as wide coverage, low communication latency, high transmission capacity, and minimal regional impact, which can provide ubiquitous internet access services. However, the development of satellite network also faces challenges like heterogeneity between satellite and terrestrial networks, and security threats from open channels and exposed nodes. Blockchain technology, with its characteristics of decentralization, distributed ledgers, immutability, and strong data consistency [1], can create a secure and trustworthy transaction environment. Applying blockchain technology in satellite network for evidence traceability, data sharing, business collaboration, and trust management can build a satellite-terrestrial integrated blockchain system. This would enable better control of satellite data, improve service guarantees, and establish trust mode.

There are several advantages of introducing blockchain technology into satellite network. Firstly, blockchain's distributed structure complements satellite network. Secondly, blockchain technology is used for monitoring, recording, and controlling events, thereby improving the security of satellite communication and the ability to share data among satellites and ground stations. Thirdly, blockchain can be used for autonomous satellite constellation control. Fourthly, the traceability and resistance

S. Jia · J. Chen (✉)
State Key Laboratory of Network and Exchange, Beijing University of Posts and Telecommunications, Beijing, China
e-mail: chenjianjun@bupt.edu.cn

H. Kang
China Satellite Network System Co., Ltd., Beijing, China

X. Jia et al. (eds.), *2nd International Conference on Big Data, IoT, and Cloud Computing*, EAI/Springer Innovations in Communication and Computing,
https://doi.org/10.1007/978-3-031-93570-1_1

to tampering of blockchain prevent the collapse of the entire satellite network due to a single satellite node failure. Finally, the blockchain for satellite-terrestrial integrated network ensures data immutability and eliminates the possibility of data leakage through cryptographic algorithms, decentralized verification mechanisms, and privacy protection technologies.

Currently, a lot of research work related to satellite and blockchain is being conducted. Considering the large-scale satellite network architecture, Beaujardiere analyzed the potential role, capabilities, and value of blockchain [2], laying the foundation for the integration of blockchain and satellite network. Li proposed an architecture for authentication and privacy protection and achieved communication, registration, authentication, and revocation of information [3]. Blockchain was also being researched for its application in data sharing scenarios for satellite remote sensing data. A preliminary design and architecture for blockchain-based satellite remote sensing data sharing was proposed by Liu [4]. Zhang proposed the use of edge computing to alleviate the processing problems of large amounts of remote sensing data [5]. Considering the high dynamic characteristics of the LEO communication system's network topology, Zhang proposed a new blockchain consensus mechanism to improve the fault tolerance of public blockchains in LEO communication scenarios [6]. Lu focused on efficient, fair, and transparent transactions as well as user privacy requirements, proposing a blockchain-based Remote Sensing Cloud Services (RSCS) transaction model to protect user privacy, and proposing an Algorithm Service (AS) transaction model combined with smart contracts to improve work efficiency [7].

The existing research on blockchain for satellite-terrestrial integrated network is lacking in understanding the technical architecture. The current blockchain technology is not suitable for complex and heterogeneous satellite-terrestrial integrated network. Therefore, this paper innovatively proposes an adaptive blockchain technical architecture scheme in the context of satellite-terrestrial integrated scenarios. And a system design scheme of software prototype is proposed to meet the modular design requirements. Under the condition that computing and storage resources are limited in satellite nodes, experiments were designed to show the resource utilization of the proposed architecture. A series of test results indicate that the lightweight of RAM and CPU consumption is the future development direction for blockchain applied in satellite-terrestrial integrated network.

2 Technical Architecture of Blockchain for Satellite-Terrestrial Integrated Network

2.1 Overall Technical Architecture

The current technical architecture of traditional blockchain is not adapted to satellite network. Therefore, we propose the eight-layer and two-vertical technical architecture based on the actual situation of satellite-borne resources and blockchain deployment, as shown in Fig. 1. The technical architecture has four basic layers, hardware resource layer, resource integration layer, data layer, and network layer. Besides, according to the characteristics of blockchain, there are four more layers, consensus layer, contract layer, interface layer, and application layer. At the same time, the cryptographic technology and operation and management as important parts of the design and operation of blockchain run through the hierarchical architecture. Based on this, we analyze the function of each layer. Red represents innovative technologies of blockchain for satellite-terrestrial integrated network.

2.2 *Eight Layers of Technical Architecture*

1 Hardware Resource Layer. The hardware resource layer includes satellites, ground communication stations, and ground control center that provide computing, storage, and network resources, as well as Trusted Execution Environments (TEE). This layer defines the availability of computing, storage, and network resources in the satellite-terrestrial integrated network and provides support for TEE as the bottom layer.
2 Resource Integration Layer. The resource integration layer manages and utilizes computing, storage, and network resources by using collaborative technologies of software and hardware, operating systems, and virtual machine management technologies.
3 Data Layer. The data layer is responsible for storing and managing data in the system. In blockchain technology, data is stored in blocks. So, this layer includes block structures, storage strategies for block data, and necessary database technology.
4 Network Layer. In the network layer, blockchain technology relies on a decentralized and distributed P2P network architecture to improve network security and reliability.
5 Consensus Layer. The consensus layer is vital in a blockchain system, mainly includes various consensus algorithms. These algorithms determine how transactions are processed among multiple participants. Consensus is essential for all nodes in a blockchain system to ensure security, consistency, and the accuracy of ledger data.
6 Contract Layer. The contract layer is used to encapsulate various script codes, algorithms, and more complex smart contracts generated by the blockchain system. Satellites can achieve trusted, stable, and autonomous operations through smart contracts.
7 Interface Layer. The interface layer is mainly used to encapsulate modules and provide a simple way of calling for the application layer. The application layer uses Remote Procedure Call (RPC) interfaces to communicate with other nodes and performs operations on local ledger data through SDK toolkits.

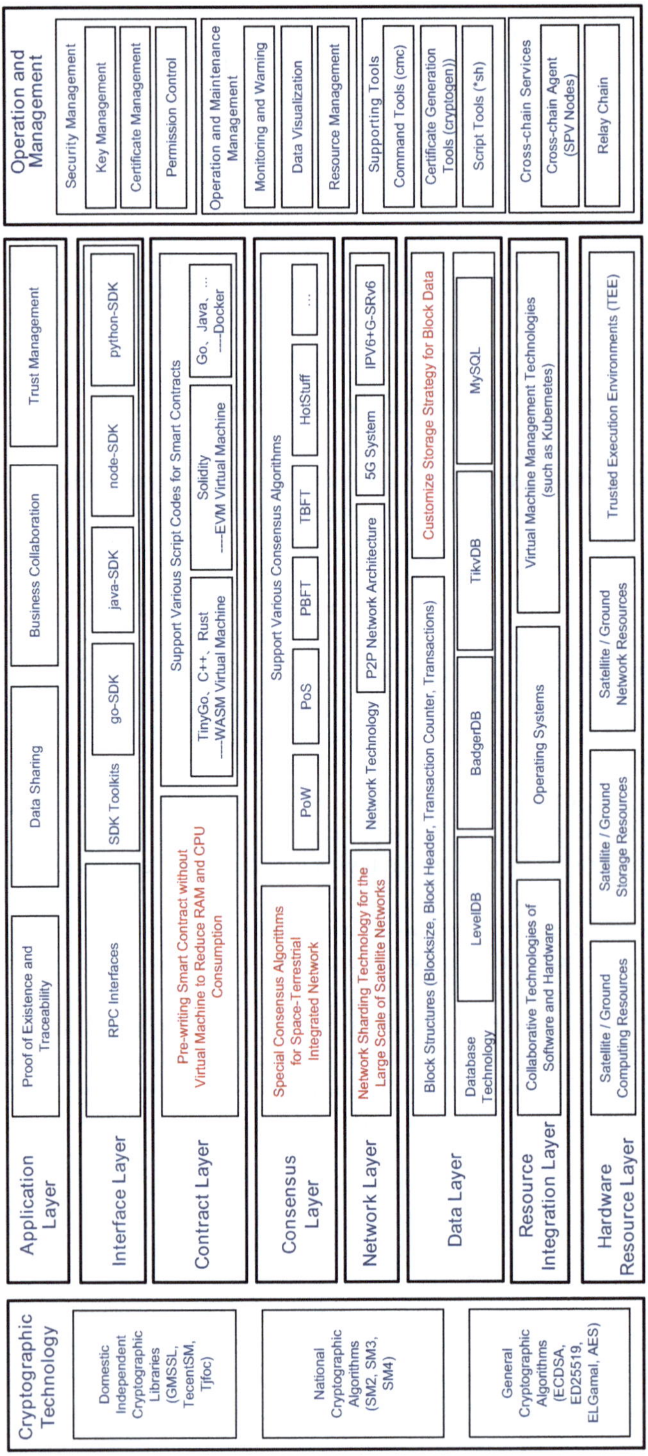

Fig. 1 A technical architecture of blockchain for satellite-terrestrial integrated network

8 Application Layer. Taking into account the characteristics of blockchain, we have identified four main application scenarios. They are proof of existence and traceability, data sharing, business collaboration, and trust management.

2.3 *Two Verticals of Technical Architecture*

1. Cryptographic Technology. Cryptographic technology is an indispensable part of a blockchain system, providing a trust foundation for decentralized information interaction and ensuring the security. Domestic independent cryptographic libraries, national cryptographic algorithms, and general cryptographic algorithms can support various application scenarios.
2. Operation and Management. Operation and management can be divided into four parts, security management, operation and maintenance management, supporting tools, and cross-chain services. Security management includes key management, certificate management, and permission control. Operation and maintenance management involves monitoring and warning, data visualization, and resource management. Supporting tools include various command tools, certificate generation tools, and script tools. Cross-chain services, as an extension of blockchain technology, ensure the security, scalability, and reliability of cross-chain transactions and enable trusted intercommunication between homogenous and heterogeneous chains, promoting the practical application development of blockchain for satellite-terrestrial integrated network.

2.4 *Superiority of the Proposed Architecture*

The differential technical module is highlighted in red in Fig. 1. We begin by analyzing the characteristics of satellite nodes and satellite-terrestrial integrated networks. Subsequently, we identify potential deployment challenges for blockchain in satellite-terrestrial integrated network and suggest an innovative technical architecture design.

The satellite-terrestrial integrated network is characterized by its complexity, heterogeneity, large number of nodes, and highly dynamic topologies. Traditional blockchain consensus mechanisms exhibit high complexity and require significant network communication resources. Therefore, when implementing blockchain technology in the satellite-terrestrial integrated network, it is necessary to shard a large number of satellite and ground nodes to reduce the burden on each node and improve system scalability. Moreover, enhancing the consensus algorithm not only reduces consensus latency but also boosts network throughput, conserves network communication resources, and ensures communication reliability. Compared to traditional blockchain, the proposed architecture utilizes network sharding technology at the Network Layer and improves the consensus algorithm at the Consensus Layer.

Satellite nodes have limited onboard resources, while traditional blockchains require significant computing and storage resources for deployment and operation. Therefore, when deploying blockchain nodes on satellites, it is necessary to apply lightweight technology. Compared to traditional blockchain, the proposed architecture envisions a series of lightweight technical for blockchain, for example, pre-writing smart contracts at the Contract Layer, improving the consensus algorithm at the Consensus Layer, and customizing the storage strategy at the Data Layer.

3 System Design Scheme of Prototype Software

We propose a system design scheme of blockchain software prototype for satellite-terrestrial integrated network, as shown in Fig. 2. In order to meet the modular design requirements of the blockchain system, we design a system supporting the flexible combination and selection of different function modules. This system supports various cryptographic algorithms to provide authentication, authorization, and privacy protection for chain members and nodes. In the software prototype system, there are three parts: on-chain management subsystem, cross-chain subsystem, and underlying-chain subsystem.

On-chain management subsystem includes the front-end module and the back-end module. The front-end module primarily presents a visual interface for man-

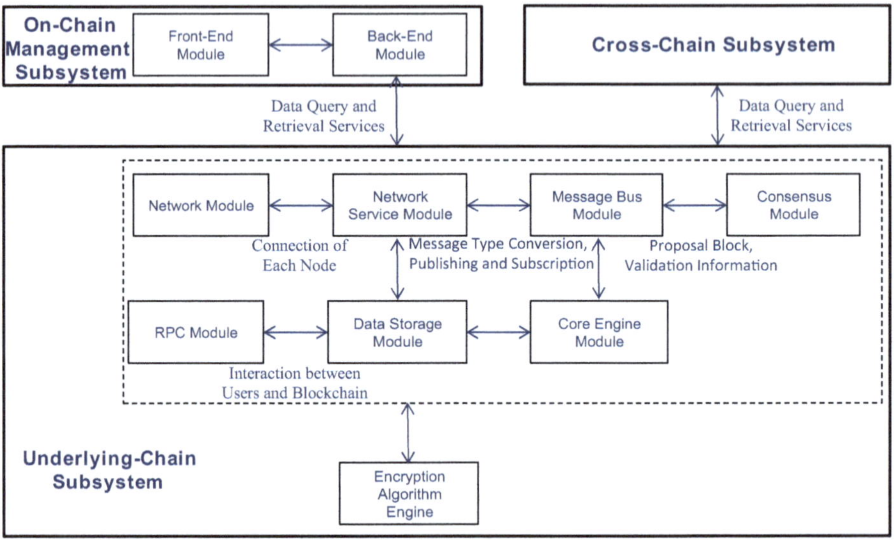

Fig. 2 A system design scheme of software prototype

aging blockchain. The back-end module includes the management back-end and SQL database. Cross-chain subsystem provides cross-chain interaction services. Underlying-chain subsystem supports message bus function, core engine function, network routing function, consensus synchronization function, data storage function, RPC function, and encryption algorithm function.

4 Results and Discussion

4.1 Experimental Conditions

We have developed a prototype software based on the eight-layer and two-vertical technical architecture, using Chain-maker as a foundation. This software is to simulate scenarios for satellite remote control and telemetry. In order to simulate the limited resources of satellite (computation, storage, and network resources), we have built a TBFT blockchain with four consensus nodes. The configuration details are provided in Table 1.

Transactions are submitted to four consensus nodes simultaneously in the blockchain. The throughput of the chain is controllable and variable and can be calculated using the following formula.

$$\text{TPS} = Nd \times Nt \times (1000/T) \tag{1}$$

where TPS is the chain-throughput; N_d is the number of nodes; N_t is the number of threads sending instructions; T is the delay in sending instructions.

To simulate satellite telemetry, tracking and control, control commands of the satellite require triple redundancy processing. They also need to be packaged with a 6-byte frame header and a 9-byte frame footer of Consultative Committee for Space Data Systems (CCSDS) standard data frame.

Table 1 Experimental parameters

Parameter name	Parameter value
Number of consensus nodes	4
Consensus strategy	TBFT
Cryptographic algorithm	TjfocSM
Block maximum capacity	400 transactions
Default outgoing block interval	200 m
Smart contract execution environment	WASMER
Database type	LevelDB

Table 2 CPU utilization rate and RAM footprint with 8 bytes commands

Chain-throughput/TPS	20	40	60
CPU utilization rate/%	3.93	5.15	6.39
RAM footprint/MB	380	400	480
Upstream bandwidth/mbps	0.91	1.32	1.82
Downstream bandwidth/Mbps	0.91	1.32	1.82

Table 3 CPU utilization rate and RAM footprint with 32 bytes commands

Chain-throughput/TPS	20	40	60
CPU utilization rate/%	3.96	5.18	6.56
RAM footprint/MB	Overflow	Overflow	Overflow
Upstream bandwidth/mbps	0.99	1.34	2.04
Downstream bandwidth/mbps	0.99	1.34	2.04

Table 4 CPU utilization rate and RAM footprint with 128 bytes commands

Chain-throughput/TPS	20	40	60
CPU utilization rate/%	3.90	5.28	6.53
RAM footprint/MB	Overflow	Overflow	Overflow
Upstream bandwidth/mbps	1.10	1.44	2.25
Downstream bandwidth/mbps	1.10	1.44	2.25

4.2 Experimental Results

Under the same testing conditions and limited node resources, with the blockchain restricting the single node CPU cores to 2 cores and RAM to 512 MB, we conducted on-chain deposition tests for satellite control commands of three commonly used lengths (8 bytes, 32 bytes, and 128 bytes). The chain-throughput for the tests was 20 TPS, 40 TPS, and 60 TPS, respectively. The results are shown in Table 2, Table 3, and Table 4.

4.3 Discussion

Experimental results validate the feasibility and effectiveness of the proposed technology architecture and software prototype. Based on the experimental results, the phenomena and causes are analyzed as follows.

1. The CPU utilization rate increases with the increase of chain-throughput. This is due to the fact that higher chain-throughput requires a single node to process more transactions, resulting in increased computational resource consumption.
2. The RAM footprint increases with the increase of chain-throughput. When the instruction length is 8 bytes and the throughput is 60 TPS, the RAM footprint reaches 480 MB, approaching the upper limit of satellite-borne RAM resources. Additionally, tests with chain-throughput of 20 TPS, 40 TPS, and 60 TPS, will result in RAM overflow and node crashes, when the commands' length is 32 bytes or 128 bytes.

3. The upstream and downstream bandwidth also increase with the increase of chain-throughput. This is because the higher the chain-throughput, the larger the transaction processing volume per unit of time, resulting in a higher data volume being transmitted per unit of time.
4. Through a series of tests, we found that the satellite-borne RAM would experience overflow first when the CPU core count is 2 and the memory is 512 MB. The CPU utilization would also be close to the upper limit. The priority order could be defined as RAM, CPU, and bandwidth. Therefore, in order to utilize blockchain in satellite network, we will research the lightweight of RAM and CPU consumption in the future. The execution environment of smart contracts is usually virtual machines, which greatly consume computing resources. We could pre-writing smart contracts to eliminate the use of virtual machines, which effectively reduces CPU and RAM consumption.

5 Conclusion

This paper proposes an eight-layer and two-vertical technical architecture of blockchain for satellite-terrestrial network and designs a system scheme of prototype software. Besides, experiments were implemented to test the CPU utilization rate, RAM footprint, upstream bandwidth, and downstream bandwidth under the condition of limited resources of satellite blockchain nodes. It was shown that the priority order is RAM, CPU, and bandwidth resources in satellite blockchain nodes. In future research, the most important thing is to reduce the RAM and CPU consumption. This is crucial for the practical application of blockchain on satellites.

Acknowledgments This work was supported by "S&T Program of Hebei (23310301D)" and "the Fundamental Research Funds for the Central Universities (2023RC56)."

References

1. Johar, S., Ahmad, N., Asher, W., Cruickshank, H., Durrani, A.: Research and applied perspective to Blockchain technology: a comprehensive survey [J]. Appl. Sci. **11**(14), 1–42 (2021)
2. Beaujardiere, J.D.L., Mital, R., Mital, R.: Blockchain application within a multi-sensor satellite architecture [C]. In: IGARSS 2019-2019 IEEE International Geoscience and Remote Sensing Symposium, pp. 5293–5296. IEEE (2019)
3. Li, C., Sun, X., Zhang, Z.: Effective methods and performance analysis of a satellite network security mechanism based on blockchain technology. IEEE Access. **9**, 113558–113565 (2021)
4. Liu H, Wang G, Li W Preliminary discussion of blockchain application in satellite remote sensing data share [C]. 2022 2nd International Conference on Computer Science and Blockchain (CCSB). IEEE, 2022:53–57
5. Zhang, L., Gao, Y., Chen, J., et al.: Research on remote sensing data sharing model based on Blockchain technology [C]. In: Proceedings of the 2019 2nd International Conference on Blockchain Technology and Applications, pp. 59–63 (2019)

6. Zhang, Z., Guo, B., Zhu, L., et al.: A public Blockchain consensus mechanism for fault-tolerant distributed computing in LEO satellite communications. China Commun. **19**(7), 110–123 (2022)
7. Lu, Y., Wang, X., Wei, R., et al.: Design of transaction system for remote sensing cloud service based on blockchain [C]. In: Proceedings of the 2019 2nd International Conference on Blockchain Technology and Applications, pp. 81–85 (2019)

Minimum Function Set of Blockchain in Satellite Network

Xinyue Wang, Hailong Kang, and Jianjun Chen

1 Introduction

Satellite network is a comprehensive service constellation that integrates wide-area communication, information gathering, navigation enhancement, and scientific experiments. It has characteristics such as wide coverage, low latency, broadband, and low cost. The data is at risk of unauthorized user tampering and eavesdropping due to the heterogeneity and wireless channel vulnerability of satellite network. Additionally, the security measures for communication in satellite network are not yet mature. Blockchain technology has features such as distributed ledger, resistance to tampering, and strong data consistency. It can be used for trustworthy data certification and ensure the objectivity, authenticity, and usability of data in satellite network. Introducing blockchain technology can be of great research significance in solving the security and reliability problems faced by satellite network.

The combination of blockchain and satellite network can provide solutions to the problems faced by satellite network, such as inadequate network security, unreliable services, and insecure sharing. Most blockchain platforms have been built not only the core framework but also a variety of component libraries and toolkits to support multiple application scenarios. The huge deployment code of blockchain is caused, and large resource space is needed by nodes. Nevertheless, satellite nodes have limited resource space and cannot bear too much resources. Therefore, it is urgent

X. Wang · J. Chen (✉)
State Key Laboratory of Network and Exchange, Beijing University of Posts and Telecommunications, Beijing, China
e-mail: chenjianjun@bupt.edu.cn

H. Kang
China Satellite Network System Co., Ltd., Beijing, China

X. Jia et al. (eds.), *2nd International Conference on Big Data, IoT, and Cloud Computing*, EAI/Springer Innovations in Communication and Computing,
https://doi.org/10.1007/978-3-031-93570-1_2

to carry out deployment lightweight of blockchain in satellite network to solve the aforementioned problem.

There are currently several blockchain deployment solutions for resource-constrained scenarios, including Tikiri [1], LightChain [2], and DewBlock [3]. These solutions are all dedicated to providing lightweight and efficient blockchain capabilities in resource-constrained environments. Among them, consensus is achieved by Tikiri using Apache Kafka, and real-time transactions are handled through a distributed system architecture of microservices. A fragmentation-based data replication method is selected to reduce the performance overhead of communication and computation. LightChain is a lightweight blockchain system designed for Industrial Internet of Things (IIoT). The collaboration of IIoT devices is incentivized through the adoption of a lightweight consensus mechanism called "collaborative multi proof." The performance is optimized by utilizing a lightweight data structure called LightBlock and an unrelated block unloading filter. DewBlock is a blockchain platform based on the Cloud-Dew architecture [4, 5]. The storage and computation of the blockchain are centralized on cloud servers to alleviate the computational and storage pressure on individual devices. However, this also brings about concerns such as privacy leakage and data security.

From the above references, it can be seen that existing lightweight deployment solutions for blockchain mainly target ground-based Internet of Things (IoT) scenarios. However, these solutions can bring about problems such as increased network overhead, privacy leakage, and data security. In addition, the functionalities of blockchain in ground network are being enriched through continuous version iterations. However, some additional functionalities may not have practical value in satellite network due to the heterogeneity between satellite and ground networks. Therefore, specific improvements and adaptive adjustments are needed to meet the unique requirements and limitations of satellite network. According to the lightweight requirements of blockchain in satellite network, the minimum function set is innovatively proposed to run blockchain in satellite nodes in this paper. The original code is optimized and adjusted by comparing and analyzing the core modules of blockchain programs. After system simulation tests were conducted, the simulation results show that the storage space occupied by the system is minimized while ensuring that its overall functionality and performance meet the requirements. Reference is provided for future blockchain in satellite network with limited resources.

2 Deployment Lightweight of Blockchain in Satellite Network

2.1 Overview

It is necessary to optimize and adjust the open-source program of Chainmaker in order to address the issue of limited satellite node resources in blockchain of satellite network scenario as shown in Fig. 1. The main directions for optimization

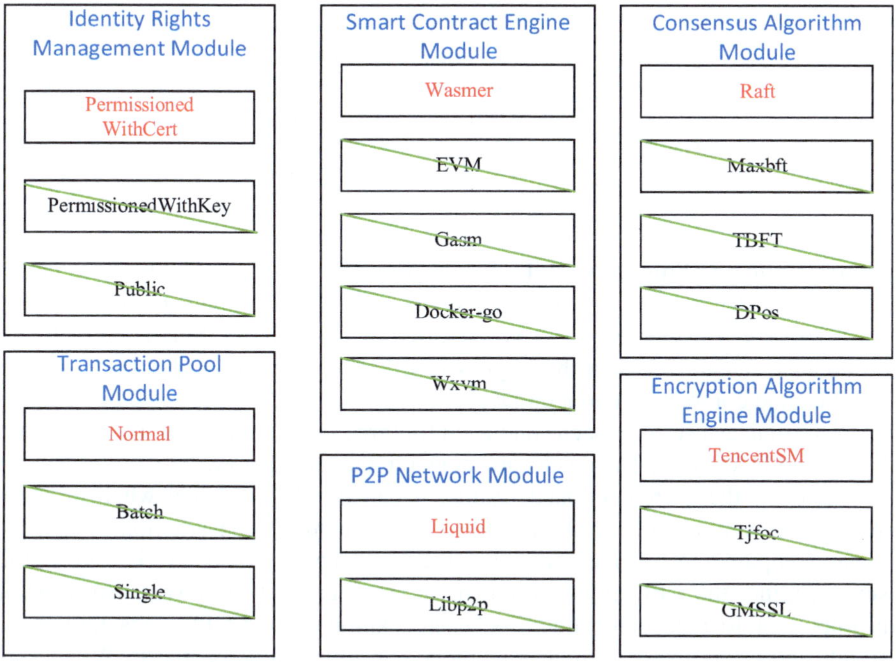

Fig. 1 Minimal functional set

are comprised of the identity rights management module, smart contract engine module, consensus algorithm module, transaction pool module, Point-to-Point (P2P) network module, and cryptographic algorithm engine module, along with other core modules. The suitable modules for the scenario need to be compared and analyzed, and the redundant functional modules should be removed in order to build the minimum functional set for running the blockchain on satellite nodes. It will help reduce the size of the compiled executable file and minimize the storage space occupied on the satellite.

2.2 *The Strategy for Building Minimal Functional Set*

Identity Rights Management Module The current identity rights management systems in Chainmaker include PermissionedWithCert, PermissionedWithKey, and Public. PermissionedWithCert and PermissionedWithKey are typically used for strong permission control scenarios, such as consortium chains, while Public is used for weak permission control scenarios, such as public chains. Blockchain in satellite network is of the consortium chain type and has a higher requirement for security. Therefore, PermissionedWithCert is adopted by minimum functional set, and the other two identity rights management modules will be removed.

Table 1 Smart contract engine performance comparison

Smart contract engine	Consensus algorithm	Encryption algorithm	CPU consumption	Memory consumption
EVM	Gmssl	TBFT	1–2 cores	800–900 MB
Wasmer	Gmssl	TBFT	1–2 cores	700–800 MB
Docker-go	Gmssl	TBFT	0.7–1.2 cores	1.2–1.3 GB

Smart Contract Engine Module Smart contracts are a set of "dynamic code" that runs on Chainmaker. At present, smart contract development using Golang, Rust, Solidity, TinyGo, and C++ is supported by Chainmaker. Smart contracts can be broadly categorized into three types based on the runtime types used: Wasm, Solidity, and contracts running on Docker.

Performance tests have been conducted for three smart contract engines: Wasmer, EVM, and Docker-go. The results are shown in the table below.

Based on the aforementioned results, it can be observed that when utilizing Wasmer as smart contract engine, the system memory consumption is notably lower compared to other engines. Conversely, when utilizing Docker-go as the smart engine, the CPU consumption of the system is at its lowest. However, it is important to note that while Docker-go exhibits lower CPU consumption, it significantly contributes to higher memory consumption compared to Wasmer as shown in Table 1.

Accordingly, Wasmer is adopted by minimum function set according to conducting a comprehensive comparison of the system resource consumption under different smart contract engines and combining the requirements for deploying smart contracts on the satellite.

Consensus Algorithm Module The consensus algorithm refers to a distributed algorithm where multiple nodes run to achieve the same data state in a distributed scenario. Currently, three common consensus algorithm types are supported by Chainmaker, namely Raft, TBFT, and Maxbft (HotStuff implementation). The performance comparison was conducted according to the characteristics and use cases of these three consensus algorithms as shown in Table 2.

The primary performance metric of concern is the computational resource consumption. Multiple rounds of voting are not required to resolve malicious node situations in the RAFT consensus algorithm, as it is a non-Byzantine fault-tolerant algorithm. Although the algorithm may face byzantine failures, it can be avoided through strict identity admission mechanisms, permission management, and other means in consortium chains, preventing the occurrence of malicious scenarios. The key point is that the algorithm can bring lower resource consumption and network bandwidth consumption compared to Byzantine fault-tolerant consensus algorithms. It still provides high fault tolerance, security, efficiency, decentralization, and resistance to double-spending.

And Raft algorithm is a distributed consensus algorithm based on the strong leader model. If we assume that there are S nodes in the distributed system, namely

Table 2 Consensus algorithm performance comparison

Algorithm	Byzantine fault tolerance	Fault tolerance	Safety	Efficiency	Consumption of computing power	Degree of decentralization	Resist to double-spending	TPS
TBFT	Yes	33%	Higher	Higher	Higher	Middle	High	200–1000
HotStuff	Yes	33%	High	High	High	High	High	400
Raft	No	50%	High	High	Low	High	High	700

Node_1, $\text{Node}_{2,} \ldots \text{Node}_s$, and within a certain time frame, the cluster receives n customer requests, namely R_1, $R_{2,} \ldots R_n$. In this case, Node_1 is designated as the leader node while the remaining nodes are follower nodes. M_{request} represents the size of the request message communication between nodes, M_{rsponse} denotes the size of the response message from a node to a client, and M_{ack} signifies the size of the confirmed message exchanged between nodes. The size of the total message sent by the leader node Node_1 in the cluster can be expressed as formula (1).

$$M_{\text{Node}_1} = \sum_1^s \text{Node}_i \times M_{\text{request}} + \sum_1^n R_i \times M_{\text{response}} + \sum_1^n R_i \times \sum_1^s \text{Node}_i \times \text{M}_{\text{request}} + \sum_1^n R_i \times \sum_1^s \text{Node}_\text{i} \times \text{M}_{\text{ack}} \tag{1}$$

In the aforementioned formula, $\sum_1^s \text{Node}_1 \times M_{\text{request}}$ represents the size of the heartbeat message sent by the leader node to all the follower nodes. $\sum_1^n R_i \times M_{\text{response}}$ represents the size of the reply to the client request message. $\sum_1^n R_i \times \sum_1^s \text{Node}_i \times M_{\text{request}}$ represents the size of the leader node sending a log replication request message to all follower nodes. $\sum_1^n R_i \times \sum_1^s \text{Node}_i \times M_{\text{ack}}$ represents the size of the leader node sending a log submission request confirmation message to all follower nodes.

The total message size sent by the follower nodes $\text{Node}_{2,} \ldots \text{Node}_s$ in the cluster can be expressed as formula (2).

$$M_{\text{Node}_{2\ldots s}} = \sum_1^n R_i \div \sum_1^s \text{Node}_i \times M_{\text{request}} + \sum_1^n R_i \div \sum_1^s \text{Node}_i \times M_{\text{ack}} \tag{2}$$

In the aforementioned formula, $\sum_1^n R_i \div \sum_1^s \text{Node}_i \times M_{\text{request}}$ represents the size of the message that the follower node receives a request from the client and then forward to the leader. $\sum_1^n R_i \div \sum_1^s \text{Node}_i \times M_{\text{ack}}$ represents the message size of the follower node responding to the leader node.

By comparing formula (1) and (2), it can be concluded that M_{Node_1} is greater than $M_{\text{Node}_{2\ldots s}}$, indicating that the leader node sends significantly larger messages than the follower nodes. Therefore, the follower nodes can process these messages with lower resource consumption.

In conclusion, RAFT consensus algorithm is adopted by minimum function set, and all other supported consensus algorithms are eliminated.

Transaction Pool Module The characteristics of the three transaction pools are compared as follows, namely single, normal, and batch as shown in Table 3.

The single transaction pool is the most basic type. System performance can be slowed down when dealing with complex transactions through serial processing of transactions. Although the batch transaction pool has the best performance, parallel batch processing of transactions increases resource consumption due to a

Table 3 Transaction pool comparison

Type	Features		
Single	Serial processing of transaction	Basic configuration	Supporting random function elimination
Normal	Parallel processing of transaction	High performance	Supporting random function elimination
Batch	Parallel batch processing of transaction	Highest performance	No supporting random function culling

Table 4 Throughput performance comparison of three cryptographic algorithm engines in the case of SM2

Password library	Sign	Verification	Encryption	Decryption
Tjfoc	3194TPS	691TPS	680TPS	984TPS
Gmssl	32793TPS	17258TPS	7633TPS	8902TPS
TencentSM	43951TPS	26937TPS	29283TPS	17054TPS

large number of concurrent transactions. Therefore, normal transaction pool type is adopted by minimal functional set, and the other two transaction pool types are eliminated.

P2P Network Module P2P network is one of the fundamental infrastructures of blockchain. P2P network model is used by blockchain as its communication method to achieve decentralized information transmission and consensus mechanism. Currently, two P2P network modules are supported by Chainmaker, namely Libp2p and Liquid. Libp2p is an open-source network protocol stack developed by Protocol Labs. Liquid is a self-developed P2P network module specifically designed to meet the network application requirements of Chainmaker, as there are some areas in which Libp2p does not fully comply with. Multi-stream multiplexing and high concurrency are supported by Liquid. It makes it well-suited for resource-constrained blockchain in satellite scenarios. Therefore, Liquid is adopted by minimal functional set, and Libp2p is eliminated.

Encryption Algorithm Engine Module Traditional signature algorithms are utilized by Chainmaker, along with a certificate system and TLS protocol. It provides reliability and basic privacy protection for communication on the blockchain. When authenticating and authorizing nodes and users, a certificate system is used. Information such as the organization and identity of nodes and users is written into the certificates, which are then signed to prevent tampering. The signature algorithm supported in the certificates includes the SM2-SM3 national cryptographic signature algorithm suite. The encryption algorithm engines used are Tjfoc, Gmssl, and TencentSM. TPS (transactions per second) of these three cryptographic algorithm engines are compared under the SM2, SM3, and SM4 scenarios by consulting reference materials as shown in Tables 4, 5, and 6.

It can be seen that TencentSM has faster computational speed under the same conditions. It can be determined that TencentSM performs with lower CPU utiliza-

Table 5 Throughput performance comparison of three cryptographic algorithm engines in the case of SM3

Password library	Hash
Tjfoc	1767591TPS
Gmssl	2159385TPS
TencentSM	3048235TPS

Table 6 Throughput performance comparison of three cryptographic algorithm engines in the case of SM5

Password library	Encryption	Decryption
Tjfoc	722229TPS	1844478TPS
Gmssl	399276TPS	484226TPS
TencentSM	768583TPS	2777679TPS

tion in the same chain running situation according to comparing the performance of different cryptographic algorithm libraries in blockchain. Therefore, TencentSM is adopted by minimal functional set, and all other supported cryptographic algorithm engines are eliminated.

3 Results and Discussion

3.1 Localization Test Environment

The domestic test environment is built with one server (processor: FT S2500), five desktop computers (processor: FT-D2000), and one laptop (processor: FT-D2000) as test terminals. The main deployment content of the server is comprised of testing auxiliary tools, namely BaaS chain management platform, resource monitoring platform, cross-chain services, and data archiving center. The four desktop computers are deployed with related node files of the test chain. Another desktop computer is responsible for browsing the front-end website of the resource monitoring platform. The laptop is used as a test terminal for using the CMC tool, calling the SDK, and writing code.

3.2 Test Results

Blockchain were run after deploying the trimmed executable file on the test chain. Data with a size of 64 bytes was stored by using a smart contract on the chain. The aforementioned data was used as the payload for test transactions.1000 transactions per second were submitted by invoking a stress-testing script in order to query the content of the on-chain data. The stress test was conducted for approximately 5 min. It was observed that there are 200,000 test transactions on-chain, with a total of 727 blocks by blockchain management platform. On average, each block includes around 250 transactions, and the size of each block is approximately 250 kB.

Table 7 Functional test sheet

Function	Functional description	Result
Service invocation	Supporting fast querying of on-chain transaction data and block state information, as well as supporting information synchronization and service invocation between anchor nodes within and outside the organization domain	Pass
Data consistency	Supporting for rapid dissemination of information and acceptance of new blocks to ensure data consistency between nodes	Pass
Node joining and exiting	Supporting the joining and exiting of new nodes and the capabilities of node discovery, isolation, and updating local information	Pass
Cross-chain capabilities	Supporting cross-chain consensus and synchronization between different organization domains, enabling controlled and fast communication between nodes in different domains	Pass

Table 8 Deployment system minimum feature set effect

Executable file type	Executable file size
Before cropping	104 MB
After cropping	72 MB

It is demonstrated that the minimal function set can run stably through testing of functions such as service invocation, data consistency, node joining and exiting, and cross-chain capabilities as shown in Table 7.

Resource monitoring platform had been used to monitor the size of executable files before and after the trimming process as shown in Table 8.

The trimmed executable file was reduced from 104 MB to 72 MB, with a reduction of 33.33%. It is demonstrated that the disk storage space can be effectively reduced by deploying the minimal functional set on blockchain in satellite network.

3.3 *Discussion*

The suitable modules for the scenario have be compared and analyzed, namely the identity rights management, smart contract engine, consensus algorithm, transaction pool, P2P network, and cryptographic algorithm engine. The redundant functional modules have been removed in order to build the minimum functional set for running the blockchain on satellite nodes. As a result, the size of the compiled executable file is reduced, thereby reducing the storage space occupation in satellite network.

The core functional modules of the Chainmaker adopt a plug-and-play design and provide multiple implementations for the same functionality. When building the minimal functional set, only redundant modules in the source code are removed. This results in a smaller compiled executable file without affecting the system's performance. Therefore, there is no impact on the system's CPU utilization and RAM usage after building the minimal functional set.

However, there are still some shortcomings in the approach we proposed. For example, certain functional modules that offer multiple techniques or methods have been removed, resulting in a lack of flexibility in online selection. However, satellite missions are usually pre-planned and relatively fixed. Therefore, it is possible to pre-determine the appropriate functional modules that are suitable for resource-constrained situations in satellite network according to the business scenario requirements on the satellite nodes.

4 Conclusion

This paper aims to research deployment lightweight of blockchain in satellite network. The contradiction is addressed between redundant blockchain system deployment codes and limited resources of satellite network nodes. This paper proposed a method of constructing minimum functional set for blockchain running on satellite nodes. System simulation tests were conducted based on the open-source Chinese-developed Chainmaker platform. The simulation results show that deployment lightweight of blockchain in satellite network can effectively reduce the disk space occupied by nodes while satisfying data security and efficient storage access requirements, without affecting the system's CPU and memory utilization rates. Reference is provided for future blockchain in satellite network with limited resources.

In the future, the further implementation of blockchain in satellite network can be expanded through the following methods. First, the management console can be adapted to minimum functional set program. This enables effective management and control of the minimal functional set program. It can improve management and monitor capabilities, and increase fault handling efficiency, and enhance system reliability and permission management. Second, the development of lightweight blockchain-specific chips can be explored. These chips have faster processing speeds, lower power consumption, higher security, and lower manufacturing costs, which helps reduce the application barriers of blockchain technology. Through these expansions, the development of blockchain in satellite network can be further advanced, achieving more efficient, secure, and intelligent blockchain applications.

Acknowledgments This work was supported by "S&T Program of Hebei (23310301D)" and "the Fundamental Research Funds for the Central Universities (2023RC56)."

References

1. Bandara, E., Tosh, D., Foytik, P., et al.: Tikiri-towards a lightweight blockchain for IoT. Futur. Gener. Comput. Syst. **119**, 154–165 (2021)

2. Zhao, C., Wang, T., Zhang, S.: Light Block: reducing bandwidth required to synchronize blocks in Ethereum network [C]. In: 2021 International Conference on Communications, Information System and Computer Engineering (CISCE), pp. 868–874. IEEE (2021)
3. Wang, Y.W.: A blockchain system with lightweight full node based on dew computing. Internet Things. **11**, 100184 (2020)
4. Wang, Y.W.: Cloud dew architecture. J. Cloud Comput. **4**(3), 199–210 (2015)
5. Wang, Y.W.: Definition and categorization of dew computing [J]. Open J. Cloud Comput. **3**(1), 1–7 (2016)

Hyperledger Fabric-Based Decentralised Digital Management System for Operational Excellence Using IoT

Dalibor Vodenicarski and Gerald Estadieu

1 Introduction

1.1 Background

A Computerised Management System (CMS) is a digital hub solution that automates certain tasks and centralises information gathering related to business operational services of an enterprise. This software tool is also intended to facilitate maintenance management offering functionalities such as organising, tracking, and scheduling maintenance operations [1].

The introduction of Computerised Management Systems (CMS) as a software tool to support management functions faces challenges in meeting diverse user requirements across various business sectors such as organisation, operations, maintenance, finance, competency, quality assurance, and health and safety processes. To surmount these difficulties, enterprises must implement a comprehensive CMS that promotes information interdependence between all business sectors. This enables integrated operational processes and facilitates efficient allocation of services, which is critical for achieving optimal performance. Designing an optimal CMS necessitates the use of an information-driven platform that is aligned with current technological trends. Such a platform facilitates the analysis of large volumes of data, enabling businesses to generate substantial value.

Moreover, by enhancing features such as mobility, connectivity, and visualisation, CMS's traditional approach will be revamped, facilitating more efficient and effective operational services. The emerging Blockchain technology could resolve

D. Vodenicarski (✉) · G. Estadieu
Faculty of Arts and Humanities, University of Saint Joseph, Macao, China
e-mail: 201600315@usj.edu.mo

X. Jia et al. (eds.), *2nd International Conference on Big Data, IoT, and Cloud Computing*, EAI/Springer Innovations in Communication and Computing,
https://doi.org/10.1007/978-3-031-93570-1_3

these challenges by incorporating specially developed applications called Decentralised Applications (DApp) managed by Smart Contracts [2]. Smart Contracts are scripts stored on the Blockchain network that can be comprehended as stored procedures in a relational database system with a unique address. They can be triggered by addressing a transaction and executing it automatically in a predefined manner [3].

The Blockchain is a distributed ledger technology that records digital transactions in a distributed transactional database. It is secured by cryptography and governed by a consensus mechanism [4]. Blockchain technology has been recognised as the foundation for genuinely free economic transactions based on its unique technological characteristics, first explained in Satoshi Nakamoto White Paper [5].

Rapid industrialisation and utilising automatisation with information technologies indicate the arrival of a fourth industrial revolution, or Industry 4.0, where electronic devices interact, exchanging data about the state of the production, products or processes, and their state [6]. Using cyber-physical systems (CPS), the Internet of Things (IoT), cloud computing, and Blockchain technology constitute a collection of pioneering information technologies. The need for more tools is an obstacle to the progress of Industrialisation 4.0 [7]. Current trends in automatisation and enabling technologies like IoT, cloud computing, Blockchain, and CPS are recognised. Industry 4.0 has emerged as an assuring technology framework used for integrating and extending the manufacturing process to respond to information integration needs into manufacturing and operational services [8].

IoT can be helpful in many business solutions as it could provide endless possibilities to the Blockchain architects to make our everyday life easier; nevertheless, its basics must be thoroughly examined and understood. There has yet to be a consensus or exact definition of the IoT. Many authors do not call IoT technology; it encompasses various technologies with different applications and solutions in other industry sectors [9].

One notable advantage of incorporating CMS in enterprises' operations is the ability to track progress in the execution of work orders, daily logs, and develop and disseminate periodic reports. Additionally, integrated operational history databases support sophisticated data analysis capabilities that are instrumental in identifying system states and potential failures, thus facilitating the development of adequate maintenance and performance improvement strategies. CMS also enables the automation of maintenance units, streamlines the planning and execution of maintenance operations, and allows for real-time communication with relevant stakeholders [10].

The IBM Maximo is an example of CMS, enterprise-grade software for asset management, allowing users to manage assets on a single platform. IBM Maximo is a cloud-based software platform that centralises predictive maintenance and reliability processes using AI, IoT, and Analytics to optimise performance and reduce operational downtime and costs. The Maximo does not simply correspond to if/yes queries; instead, it analyses the data input from machinery and predicts when breakdowns are likely to occur. Once a machine is fixed, Maximo will keep track of the data after the service is performed, what kind of service was performed, and how the machine functions afterwards [11].

This paper proposes developing an enterprise-grade data management system for CMS to improve operational services in service-providing organisations. This proposal utilises Blockchain and IoT technologies that offer decentralised data storage and information-sharing services. The Hyperledger Fabric (HLF) protocol is employed in this project, which includes data-sharing channels to ensure confidentiality and data isolation between different channels, and ensures traceability and security of the transaction data [12].

The main objective of this research is to seek the answer to how can Blockchain-based technology and its software applications, such as Smart Contracts and DApp, facilitate and support the operational excellence of enterprise services in the form of a Computerised Maintenance System solution. The final deliverable shall be modernised maintenance operations, the so-called Smart Maintenance, which is data-driven and incorporates and analyses vast amounts of data generated from the day-to-day operations of an enterprise and its IoT devices [13].

1.2 Related Works

The Blockchain technology mitigates the risk of network attacks and a single point of failure by employing distributed peer-to-peer network nodes. Decentralised policies reduce fraud by time-stamping every single transaction, and users' information is stored in an immutable distributed ledger across the network using the Smart Contract. Blockchain eradicates processes such as settlement between multiple isolated ledgers and administrative procedures, which helps reduce the system's cost. Also, using various cryptographic operations enhances the transaction speed and level of security.

The researchers conducted several surveys using Blockchain technology and IoT, which are summarised as follows.

Liu et al. [14] proposed a Blockchain-based access control system using IoT named "fabric-IoT". This system is based on the Hyperledger Fabric Blockchain framework and attributed-based access control (ABAC). The authors of this paper confirmed in their experiments that "fabric-IoT" can maintain high throughput in a large-scale data request environment and reach consensus efficiently in a distributed system to ensure data consistency. They used three different types of Smart Contracts to achieve this. Azizi et al. [15] present a pioneering investigation that explores the application of IoT and Blockchain technology in Smart Supply. The study defines Smart Supply as a form of supply chain management that involves the provision of appropriate goods at the correct location, time, and volume, under the right circumstances. The research demonstrates how various indicators of IoT and Blockchain technology contribute to Smart Supply chain management, including tool application, component connectivity, optimal decision-making, automated information delivery, integration, innovation, and learning. These findings indicate the potential benefits of utilising smart supply chains, enabled by the integration of IoT and Blockchain technology. The study [16] observed and proposed a solution to develop a Blockchain-based IoT system to establish secure communication and create an entirely decentralised cloud computing platform for vehicles or the Internet of Vehicles (IoV) concept. The IoV represents the real-time data interconnection between vehicles and infrastructure through smart terminal devices, vehicle navigation systems, mobile communications technology, and information platforms. The paper [17] proposes an expert method for Smart Cities based on Blockchain Smart Contracts and neural networks as the deduction mechanism to decide on the sensors' optimal distribution and actions when sensors readings are out of the range. Khan et al. [18] explored IoT-Blockchain data of Industry 4.0 in the food sector as a research object.

Since IoT devices are resource-constrained; therefore, the interoperability, identity, and privacy of IoT systems need to be guaranteed, and at the same time, reducing processing time is necessary. Therefore, this paper proposes an HLF-based Blockchain network to store and retrieve data created by IoT devices from the distributed ledger.

2 Proposed Scheme and Testing

2.1 System Framework for Blockchain Network with Integrated IoT

This paper elaborates on the HLF protocol-based Blockchain framework for enterprise-grade CMS, which we named a Decentralised Digital Management System (DDMS) using IoT data for sharing and analysing.

The first layer is the physical layer which contains IoT devices based on Raspberry Pi Pico W [19], such as sensors and switches that are used to sense and perceive the surrounding environment and collect relevant data to enable us to understand our surroundings. The next layer is the one that performs network and routing management, allowing all IoT objects to be linked and communicate over Web-Broker (HiveMQ) [20], using Node-Red [21] platform as virtual wiring applications.

This layer includes networking and security protocols such as OpenSSL for general-purpose cryptography that enable communication and security management on the microcontroller side and a lightweight messaging protocol called Message Queue Telemetry Transport (MQTT) for sharing data exchanged between IoT devices which is extremely simple with a publish-subscribe architecture, designed to be automatically deployed and capable of supporting thousands of clients with a single server [22].

The framework is shown in (Fig. 1).

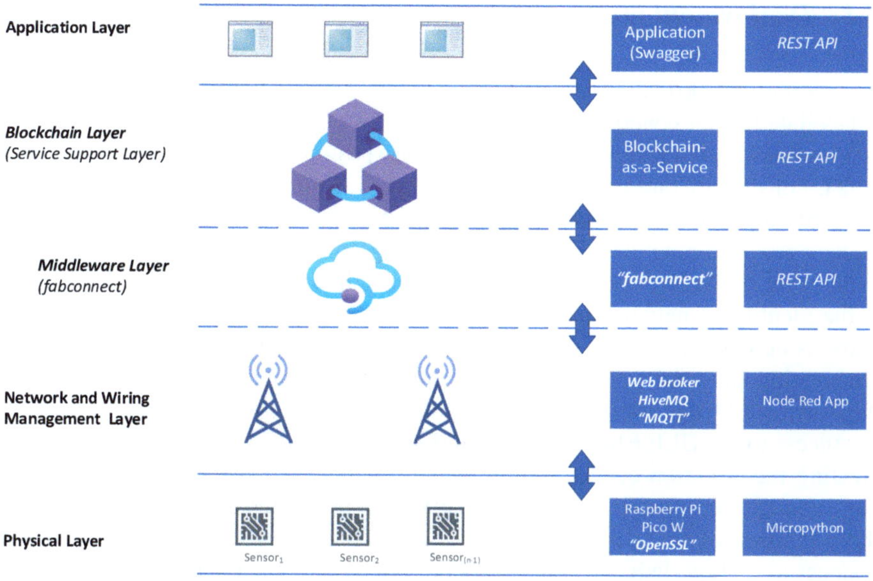

Fig. 1 The proposed five-layer blockchain framework

The newly added layer is named the fabconnect middleware layer [23], a middleware application consisting of reliable REST APIs that enable various features of the Blockchain technology to be implemented in the IoT system to interact with a Hyperledger Fabric network and stream events.

The features of the middleware layer provide three high-level sets of API endpoints that can be used for registering and enrolling identities, signing, submitting, querying for transaction results/receipts, and subscribing to events with a regex-based filter. In the Blockchain Layer or Service Support Layer, the online BaaS Web3 platform utilises the exposed REST API to receive and send transactions. It uses the network system channel to connect all peer nodes in the Blockchain network, ensuring data security and access to other services. The last application layer is used for data visualisation using DApps, a secure and transparent way for users to interact with each other without the need for intermediaries.

To evaluate the efficacy of the popular JavaScript Object Notification (JSON) [24] data format and its transactions using HLF Blockchain protocol, our research method involved creating events as transactions on a developed Blockchain network with varying conditions. We varied the frequency of transactions and network load to determine the optimal conditions for JSON data format transactions while monitoring the transaction throughput and network latency.

2.2 Deployment, Testing, and Valuation

For testing and providing the proof-of-concept for the framework that integrates JSON data format generated from IoT devices, we deployed a Blockchain network on the cloud-hosted Kaleido Web3 platform [25] using Hyperledger Fabric protocol, which consists of one system channel, Fabric Certificate Authority (Fabric-CA), two ordering nodes, and two peer nodes with installed Smart Contract written in Golang programming language (GO) for managing the event transactions. The vital component of the HLF protocol is the Membership Service Provider (MSP), which is the identity of network participants. MSP is responsible for authenticating and authorising network participants. In our case, the MSP component is crucial, as we are creating an identity outside of the network that will access and store the data on the distributed ledger through the fabconnect endpoints.

Access to the HLF Blockchain network from the local machine is established by creating the common connection profile file "ccp.yaml" and simple configuration file "config.json", which contains information about the network topology and the identities of the network participants. Hyperledger Fabric SDKs use it to connect to a Hyperledger Fabric network. Once the connection is established, the exposed REST API in the Swagger platform from the fabconnect endpoints receives from the IoT device JSON data format file and stores it on the ledger. Every transaction created from IoT device consists of its own identification number, its time stamp, and description as an identification. As a state database, we use the CouchDB NoSQL database, allowing us to use queries to retrieve stored data. Data on the CouchDB database are stored as a key-value pair. The code in the Micropython programming language creates information about the event as a JSON file, also called meta-data, and the code is uploaded to the Raspberry Pi Pico W microcontroller's internal memory (Fig. 2).

Results of benchmarking and the performance evaluation of the IoT Smart Contract for JSON data format transactions against the test network run on the local computer are completed in the Hyperledger Caliper [26] benchmarking platform. Our study confirmed that the latency of completed transactions varied depending on the complexity of the Smart Contracts, with more complex contracts, such as examples provided by the Hyperledger GitHub repositories named "fabcar" (FC) and "marbles" (MB), taking longer to process. The performance evaluation results of three Smart Contracts, including the IoT Smart Contract on a local machine with 5000 transactions, are shown in Table 1 as a comparison.

Our latency and throughput performance testing of transactions using developed IoT Smart Contract in GO for HLF Blockchain protocol showed an average transaction processing time of 0.688 s on a sample of 50 transactions created by the IoT device to our test network hosted on the cloud. Using the standard distribution sampling method to estimate transaction latency on 5000 transactions, we received results of mean transaction latency of 0.701 s with a standard deviation of 0.0485 considering the cloud-hosted Blockchain network.

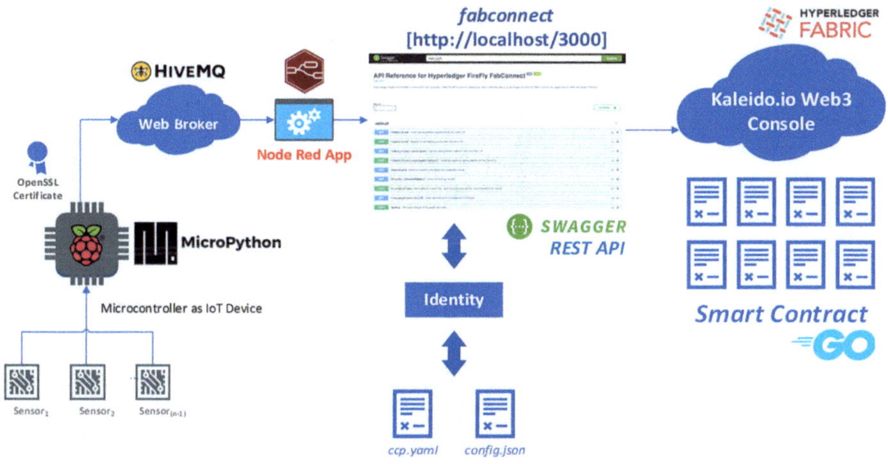

Fig. 2 Entire blockchain network, incorporating IoT device

As a final observation, all JSON meta-data transactions on the cloud-hosted Blockchain network were processed, and the block was created under 0.75 s using the exposed REST API endpoints through the fabconnect platform; the HLF protocol ensures that transactions are completed in a fast and secure manner, tamper-proof, scalable, whereas the fabconnect library provides an easy-to-use interface to interact with the HLF network hosted on the cloud. The ledger was instantly accessible with an access time of 0.00 s, enabling the final user to query it quickly and securely for various business processes.

The key contributions of our study are threefold:

1. We developed and tested a framework that enables data-sharing and privacy protection based on the Hyperledger Fabric protocol and IoT. To overcome the Blockchain load problem and achieve scalability, we utilised the cloud computing Web3 BaaS online platform.
2. Using the IoT Smart Contract, we designed a JSON format meta-data that implements a key-value pair structure to facilitate transaction management of automatic data creation, storage, and retrieval from a distributed ledger.
3. This approach enhances privacy protection for sensitive data while improving the system's scalability.

3 Conclusions

This paper provides insights into the importance of Blockchain technology and its integration with the IoT for the Decentralised Digital Management System (DDMS) and various other smart applications where security and decentralisation

Table 1 Performance evaluation results for three Smart Contracts “IoT”, “fabcar”, and “marbles”

Name		Succ	Fail	Send rate [TBS]	Max latency [s]	Min latency [s]	Average latency [s]	Throughput [TPS]
IoT	IoT event	5000	0	27.3	2.09	0.06	0.24	27.2
	Query IoT events	41,478	0	703.0	0.06	0.00	0.01	702.9
FC	Create a car.	5000	0	30.2	2.09	0.04	0.21	29.9
	Query all cars	18,917	0	652.2	0.04	0.01	0.01	652.1
MB	Init marbles	5000	0	25.0	0.33	0.06	0.19	25.0
	Query marbles	305	0	5.1	0.64	0.52	0.58	5.0

are paramount. The authors conclude that DDMS based on the Smart Maintenance and IoT paradigm could be incorporated to modernise operations to achieve Operational Excellence making all records stored on the ledger swiftly available for an edge in data analyses, compiling and computing. Blockchain-based DDMS can improve security in IoT-based systems by providing a local authentication process and traceability for the data generated by IoT devices. The proposed framework will enable interconnected sensors and subsystems to use data from other sources like an enterprise's finance or accounting, operations, health and safety, maintenance, or competency business segments. Future research must also be focused on data structure, security issues, development of Smart Contracts, limited processing power, and storage capacity when incorporating Blockchain technology in IoT, which need help to afford complex and heavyweight operations.

References

1. Bagadia, K.: Computerized Maintenance Management Systems Made Easy: How to Evaluate, Select, and Manage CMMS, 1st edn. McGraw-Hill, New York (2006)
2. Tapscott, D., Tapscott, A., Blockchain Revolution: How the Technology Behind Bitcoin and Other Cryptocurrencies Is Changing the World, First Portfolio/Penguin Trade. Paperback edn. Portfolio/Penguin, New York (2018)
3. Rakovic, V., Karamachoski, J., Atanasovski, V., Gavrilovska, L.: Blockchain paradigm and internet of things. Wirel. Pers. Commun. **106**(1), 219–235 (2019)
4. Beck, R., Michel, A., Rossi, M.: Blockchain technology in business and information systems research. Bus. Inf. Syst. Eng. **59**(6), 381–384 (2017)
5. Nakamoto, S.: Bitcoin: A Peer-to-Peer Electronic Cash System, p. 9 (2008)
6. Bashar Fakhri, A., et al.: Industry 4.0: architecture and equipment revolution. Comput. Mater. Contin. **66**(2), 1175–1194 (2021)
7. Xu, L.D., Xu, E.L., Li, L.: Industry 4.0: state of the art and future trends. Int. J. Prod. Res. **56**(8), 2941–2962 (2018). https://doi.org/10.1080/00207543.2018.1444806
8. Adaros Boye, C.A.: Understanding Cyber-Risks in IoT: When Smart Things Turn Against You. Business Expert Press, New York (2019)
9. Alshalalfah, B., Shalaby, A., Dale, S., Othman, F.M.Y.: Aerial ropeway transportation Systems in the Urban Environment: state of the art. J. Transp. Eng. **138**(3), 253–262 (2012)
10. Ogbeifun, E., Pasipatorwa, P., Pretorius, J.-H.C.: Harnessing the multiple benefits of a computerised maintenance management system. In: Petrillo, A., De Felice, F., Lambert-Torres, G., Bonaldi, E. (eds.) Operations Management—Emerging Trend in the Digital Era. IntechOpen (2021)
11. Plaskett, E.: Using AI for predictive maintenance, employee development. Mod. Mach. Shop. **92**(5), 58–60 (2019)
12. Vodenicarski, D.: How Digital Transformation Could Facilitate Operational Improvements: the Case of Urban Cable Car Industry. University of Saint Joseph, Macau (2019)
13. Bokrantz, J., Skoogh, A., Berlin, C., Wuest, T., Stahre, J.: Smart maintenance: an empirically grounded conceptualization. Int. J. Prod. Econ. **223**, 107534 (2020)
14. Liu, H., Han, D., Li, D.: Fabric-iot: a Blockchain-based access control system in IoT. IEEE Access. **8**, 18207–18218 (2020)
15. Azizi, N., Malekzadeh, H., Akhavan, P., Haass, O., Saremi, S.: IoT–Blockchain: harnessing the power of internet of thing and Blockchain for smart supply chain. Sensors. **21**(18), 6048 (2021)

16. Jabbar, R., Kharbeche, M., Al-Khalifa, K., Krichen, M., Barkaoui, K.: Blockchain for the internet of vehicles: a decentralized IoT solution for vehicles communication using ethereum. Sensors. **20**(14), 3928 (2020)
17. Aguilera, R.C., Ortiz, M.P., Ortiz, J.P., Banda, A.A.: Internet of things expert system for smart cities using the Blockchain technology. Fractals Singap. **29**(1), 2150036 (2021)
18. Khan, P.W., Byun, Y.-C., Park, N.: IoT-Blockchain enabled optimized provenance system for food industry 4.0 using advanced deep learning. Sensors. **20**(10), 2990 (2020)
19. R. P. Ltd. Raspberry Pi Pico. Raspberry Pi. https://www.raspberrypi.com/products/raspberry-pi-pico/. Accessed July 11, 2023
20. The Free Public MQTT Broker by HiveMQ—Check out our MQTT Demo. https://www.hivemq.com/public-mqtt-broker/. Accessed 29 July 2023
21. Node-RED. https://nodered.org/. Accessed Jul. 13, 2023
22. Hernández Ramos, S., Villalba, M.T., Lacuesta, R.: MQTT security: a novel fuzzing approach. Wirel. Commun. Mob. Comput. **2018**, 11 (2018)
23. firefly-fabconnect. Hyperledger, Jul. 06, 2023. Accessed: Jul. 11, 2023. [Online]. Available: https://github.com/hyperledger/firefly-fabconnect
24. Lv, T., Yan, P., He, W.: Survey on JSON data modelling. J. Phys. Conf. Ser. **1069**(1), 12101 (2018). https://doi.org/10.1088/1742-6596/1069/1/012101
25. Kaleido: Enterprise-Grade Blockchain & Digital Asset Platform. https://www.kaleido.io/. Accessed May 22, 2022.
26. Hyperledger Caliper. https://hyperledger.github.io/caliper/. Accessed Jul. 20, 2023

Biomedical Knowledge Graph Completion with Efficient Contrastive Learning

Jing Qian, Yong Yue, Katie Atkinson, and Gangmin Li

1 Introduction

Knowledge graphs (KGs) have emerged as an indispensable approach for integrating heterogeneous biomedical data sources, owing to their capacity to store a vast amount of expert-curated domain knowledge. The integration of diverse data sources is critical for crucial data-driven tasks such as personalized medicine and drug repurposing, which rely on the amalgamation of information from distinct biomedical knowledge graphs. These graphs encompass a wide range of biological entities and relations, including proteins, genes, drugs, diseases, and intricate interactions among them. Typically, knowledge graphs are composed of collections of triples (h, r, t), where h and t refer to the head and the tail entity, and r denotes their relationship. Some examples of well-known biomedical knowledge graphs include RepoDB [1], DrugBank [2], and Hetionet [3], among others.

Although real-world KGs contain an extensive number of triples, incompleteness remains a persistent issue. As a result, KG completion (KGC) has become a fundamental task in which unknown facts are inferred on KGs. The development of complete and refined biomedical knowledge graphs is essential in achieving more accurate outcomes in biomedical tasks. KGC methods can be broadly classified into two categories: graph embedding and textual encoding methods. Graph embedding

J. Qian · Y. Yue (✉)
School of Advanced Technology, Xi'an Jiaotong-Liverpool University, Suzhou, China
e-mail: yong.yue@xjtlu.edu.cn

K. Atkinson
Department of Computer Science, University of Liverpool, Liverpool, UK

G. Li
School of Computer Science Technology, University of Bedfordshire, Luton, UK

X. Jia et al. (eds.), *2nd International Conference on Big Data, IoT, and Cloud Computing*, EAI/Springer Innovations in Communication and Computing,
https://doi.org/10.1007/978-3-031-93570-1_4

methods refer to the family of knowledge graph embedding (KGE) models that approach the KGC task by learning low-dimensional vector representations of entities and relations. These representations are then utilized in a scoring function to estimate the likelihood of a given triple. However, most embedding-based methods can only predict missing links within observed triple facts, restricting their applicability in real-world scenarios involving dynamic graphs where the constant addition of new entities occurs. Furthermore, conventional KGE models such as TransE [4], TransH [5], DistMult [6], and ConvE [7] cannot incorporate any side information, such as textual entity descriptions.

In contrast to graph embedding methods, textual encoding methods [8–10] train entity encoders that integrate textual data available in KGs to learn entity representations. These encoders are implemented using deep neural networks such as pre-trained language models (PLMs) [11]. Recently, PLMs such as BERT [12] and RoBERTa [13] have achieved exceptional performance in various language understanding tasks, as they can learn contextualized word embeddings from abundant unlabeled text data using language modeling objectives. Although text-based methods should theoretically outperform embedding-based methods because of their capacity to fully exploit the textual information available, recent studies [14] suggest otherwise. Text-based methods, even with PLMs, still fall short of embedding-based methods. For simplicity, we call textual encoding methods with PLMs as PLM-based methods.

The primary reason for the inferior performance of textual encoding methods is attributed to the inefficiency of contrastive learning. Contrastive learning has emerged as a prevailing element within self-supervised learning methodologies, exerting its influence across diverse domains such as computer vision, natural language processing (NLP), and other related fields. It is a discriminative approach that aims to embed augmented versions of the same sample (positive sample) close to each other while pushing away dissimilar samples (negative samples). The effectiveness of contrastive learning methods in producing high-quality representations relates to the quantity of negative samples employed. Graph embedding methods do not require the computationally expensive text encoders and can be trained efficiently with a large negative sample size. In contrast, due to the substantial computational burden associated with PLMs, the performance of PLM-based methods is constrained to a limited negative sample size.

This paper takes inspiration from SimKGC [14], which employs contrastive learning with InfoNCE loss [15] in KGC task. We utilize in-domain PLMs in our ensemble methods to enhance the performance of PLM-based methods in biomedical KGC. Specifically, we combine three ways to generate contrastive samples, namely in-batch negatives [16], pre-batch negatives [17], and hard negatives [18]. In terms of model architecture, we employ a two-tower model with PLMs to encode entities and relations separately. Recent research [19, 20] has demonstrated that the parameters of PLMs can contain abundant linguistic, relational, and factual knowledge. Furthermore, these models can store domain-specific knowledge in their pre-trained weights if they undergo further domain-adaptive pre-training. Since our experiments focus on the biomedical domain, we consider four biomedical PLMs, namely BioBERT [21], SciBERT [22], PubMedBERT [23], and SapBERT [24], as candidate encoders in our ensemble methods. By comparing their performance against representative embedding-based methods and other text-based methods, we demonstrate the effectiveness of our proposed combination of in-domain PLMs and efficient contrastive learning.

To summarize, our work contributes in the following ways. Firstly, it is the first attempt to enhance PLM-based methods with efficient contrastive learning for biomedical knowledge graph completion. Secondly, we develop a two-tower model with four different biomedical PLMs and verify the effectiveness of our ensemble methods on two benchmark datasets, UMLS and Hetionet. Lastly, experimental results substantiate that our proposed framework surpasses state-of-the-art KGC methods on biomedical KGC in terms of automatic evaluation metrics.

2 Related Work

2.1 Knowledge Graph Completion

Knowledge Graph Completion (KGC), commonly known as link prediction, aims at predicting the absence of relationships between entities within a KG. Approaches to address this task can be roughly categorized into graph embedding and textual encoding methods. Embedding-based methods project nodes and edges within KGs into low-dimensional vector space, without utilizing any side information like entity descriptions. A scoring function operating in the embedding space calculates the likelihood of observed triples. Popular graph embedding methods include translational distance models like TransE [4] and TransH [5], where a triple (h, r, t) represents a relation-specific translation from the head entity h to the tail entity t. Additionally, semantic matching models such as DistMult [6] and ComplEx [25] assess the plausibility of triples by evaluating the alignment of latent semantics present in the vector space representations of entities and relations.

While graph embedding methods preserve both local and global structural information of KGs, they are not applicable in scenarios where new entities are added to the graph. In contrast, textual encoding methods train encoders to embed not only components of KGs but also additional information such as entity descriptions. DKRL [26] uses a convolutional neural network (CNN) to encode text, but its CNN architecture lags behind recent advances in neural networks for NLP, such as PLMs [11]. KG-BERT [8] treats triples in KGs as textual sequences and calculates the scoring function of the triples with BERT [12]. Likewise, KEPLER [27] harnesses BERT [12] as an entity encoder, which undergoes training with a combined objective that encompasses both language modeling and link prediction. However, text-based methods still underperform embedding-based methods like ConvE [7]. A recent work, SimKGC [14], proposes to increase the efficiency of contrastive learning to improve the performance degradation of textual encoding methods.

2.2 *Contrastive Self-supervised Learning*

Contrastive learning has emerged as a prominent self-supervised learning approach for representation learning, which aims to learn a low-dimensional embedding space where similar data samples are embedded close to each other, while dissimilar samples are separated by a large margin [15, 28]. In recent years, contrastive learning with deep neural networks has achieved remarkable success in various vision and language applications [18, 28, 29]. Several frameworks have been introduced for contrastive learning of visual representations, including SimCLR [28], which utilizes augmented views of other items within a mini-batch as negative samples, and MoCo [30], which employs a memory bank of prior negative representations to facilitate the utilization of large batches of negative samples.

Contrastive learning was originally proposed by [29] for learning word embeddings and has recently gained popularity in various NLP downstream tasks. For instance, Sentence-BERT [31] employs Siamese networks with triplet loss to acquire sentence embeddings by considering cosine similarity, CERT [32] utilizes contrastive self-supervised learning at the sentence level, incorporating back-translation data augmentation techniques. Also, negative sampling is a crucial step in knowledge representation learning, and the quality and quantity of negatives directly affect learned representations. SimKGC [14] leverages contrastive learning with PLMs to address the performance degradation of text-based methods compared to embedding-based methods in KGC. Our work aligns with this line of research, but is the first attempt to utilize contrastive learning specifically for Biomedical KGC that requires domain awareness other than general knowledge.

3 Methodology

3.1 *Notations*

A Knowledge Graph (KG) is typically composed of triples $(h, r, t) \in \mathbf{K}$, where h is the head entity, t is the tail entity ($h, t \in \mathbf{E}$), and they are connected via the relation r ($r \in \mathbf{R}$). Many real-world KGs are known to be incomplete and noisy. Therefore, multiple techniques have been developed to predict missing links based on existing facts in $\mathbf{K}$ [33]. These techniques consider partial triples such as $(h, r, ?)$ and attempt to predict the correct tail entity t, or are provided with $(?, r, t)$ and attempt to predict the correct head entity h. We refer [7] to denote the inverse relation for a given relation $r \in \mathbf{R}$ as r^{-1}, and then (h, r, t) is equivalent to (t, r^{-1}, h). Therefore, we only need to focus on tail entity prediction.

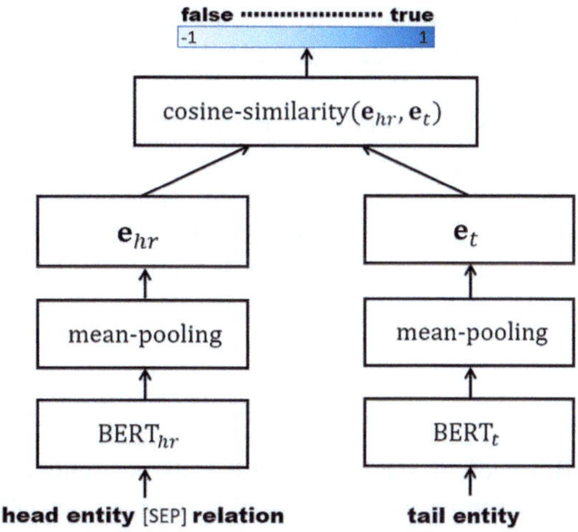

Fig. 1 The two-tower model with two BERT-structured encoders

3.2 *The Contrastive Learning Framework*

Model Architecture As illustrated in Fig. 1, our contrastive learning framework is constructed on a two-tower model. Similar to Sentence-BERT [31], our dual-encoder comprises two fully decoupled BERT-based models that are isomorphic, with one dedicated to encoding the head and relation, and the other to encoding the tail. Both encoders are initialized with the same biomedical PLM, but they do not share weights.

Given a triple (h, r, t), the left encoder, BERT_{hr}, is utilized to generate the relation-aware embedding $\mathbf{e}_{hr}$ for the head entity h. The input for BERT_{hr} is the concatenation of the textual sequences of the head entity h and the relation r, with a special symbol [SEP] inserted between them. The output $\mathbf{e}_{hr}$ is relation-aware because the same head entity with different relations will generate distinct embeddings. Similarly, the right encoder, BERT_t, is utilized to generate the embedding $\mathbf{e}_t$ for the tail entity t. And the input for BERT_t only includes the text data about the tail entity t. By performing mean-pooling over all token representations from the last hidden layer, the encoders generate separate representations for the relation-aware head entity and the tail entity, respectively.

Typically, a scoring function is formulated to assess the plausibility of a provided triple. As the embeddings $\mathbf{e}_{hr}$ and $\mathbf{e}_t$ are contextualized representations, they tend to be close in the vector representation space if they possess similar meanings. Hence, a similarity score is computed between these two embeddings using a cosine distance similarity metric.

$$\text{score} = s\,(h, r, t) = \cos\,(\mathbf{e}_{hr}, \mathbf{e}_t) = \frac{\mathbf{e}_{hr} \bullet \mathbf{e}_t}{\|\mathbf{e}_{hr}\|\ \|\mathbf{e}_t\|} \tag{1}$$

Regarding tail entity prediction for a given triple $(h, r, ?)$, the cosine similarity is calculated between $\mathbf{e}_{hr}$ and all entities in $\mathbf{E}$, and the entity with the highest score is predicted. However, knowledge graphs usually have spatial locality, meaning that nearby entities are more likely to get high scores. While textual encoding methods excel at capturing semantic relatedness, they may not handle well with such inductive bias. Therefore, we adopt a straightforward graph-based re-ranking strategy, as proposed by SimKGC [14], whereby the score of candidate entity t_i is increased by α ($\alpha \geq 0$) if t_i is among the k-hop neighbors $\mathbf{E}_k(h)$ of the head entity h. We set α to 0.05 in our work.

$$\arg\max_{t_i} s\left(h, r, t_i\right) + \alpha \mathbb{1}\left(t_i \in \mathbf{E}_k(h)\right) \tag{2}$$

Biomedical BERTs BERT [12] is architectured from bidirectional encoder of Transformer and pre-trained with objectives of masked language modeling (MLM) and next sentence prediction (NSP). Transformer [11] employs a multi-layer, multi-head self-attention mechanism, which facilitates parallel computation and captures long-range dependencies within textual data. To obtain augmented versions of PLMs tailored to specific domains, domain-specific pre-training from scratch or continual pre-training of general-domain PLMs would be considered. Recently, BERT [12] has been developed in the biomedical domain and its variants including BioBERT [21], SciBERT [22], PubMedBERT [23], and SapBERT [24] are produced. Pre-training details for these four biomedical BERTs are summarized in Table 1.

BioBERT [21] *is the first biomedical PLM, which basically has the same structure as BERT* [12]. Initially, BioBERT's weights are initialized using the standard BERT model, which is pre-trained on corpora from the general domain, including English Wikipedia and BooksCorpus. Subsequently, BioBERT undergoes additional pre-training on corpora specific to the biomedical domain, such as PubMed abstracts (PubMed) and PubMed Central full-text articles (PMC). The continual pre-training is based on the original vocabulary that is generated from Wikipedia and BookCorpus. Such expedient offers convenience, but base vocabulary fails to adequately represent the intricate terminologies and word nuances of the biomedical domain.

Table 1 Summary of pre-training details for the various BERT models used in our work

Model	Vocab	Pre-training	Corpora
BERT	General	–	Wikipedia and BooksCorpus
BioBERT	General	Continual	Wikipedia, BooksCorpus, PubMed, PMC
SciBERT	In-domain	From scratch	PMC and CS papers
PubMedBERT	In-domain	From scratch	PubMed
SapBERT	In-domain	Continual	PubMed and UMLS synonyms

SciBERT [22] *is pre-trained from scratch on the full text of 1.14 M biomedical and computer science (CS) papers.* Its configuration and size align with standard BERT model. Furthermore, SciBERT uses an in-domain vocabulary, which is different from BioBERT that uses original BERT vocabulary. The pre-training of SciBERT still belongs to mixed-domain because the inclusion of computer science text signifies content that is clearly out-domain when considering biomedical applications.

PubMedBERT [23] *Challenges the Commonly Held Assumption That Mixed-domain Pre-training is Universally Applicable.* It conducts domain-specific pre-training from scratch, i.e., pre-trained using purely biomedical text with in-domain vocabulary.

SapBERT [24] is Essentially an Extension of PubMedBERT, Wherein it Conducts Continual Pre-training on UMLS Synonyms to Obtain enhanced biomedical entity representations. It is difficult to separate biomedical names of distinct concepts in the embedded space. To address this issue, SpaBERT introduces a scalable metric learning framework that facilitates the self-alignment of the representation space for synonymous entities. Then the entity names corresponding to the same concept converge to form compact clusters, which benefits entity-level tasks such as entity linking.

Contrastive Samples. KGE models are normally trained in a discriminative way. Triples in a KG are true positive samples, and thus negative sampling is a key step in knowledge representation learning. Several sampling strategies [4, 5] have been proposed to corrupt the head or tail entities to generate contrastive samples. However, the negative sample size of graph embedding methods lags behind that of textual encoding methods [9]. This work incorporates three distinct types of negative samples to enhance the training efficiency of contrastive learning.

In-batch Negatives (IN) [16] Refers to the Rest $\boldsymbol{B} - \mathbf{1}$ *Samples within the Same Batch Excluding the Selected Positive One, for a Batch of Size* $\boldsymbol{B}$. The number of in-batch negatives is positively related with batch size; however, further increasing the batch size would result in significant overhead of GPU memory. Due to limited computational resource, we set batch size as 256.

Pre-batch Negatives (PN) [17] is a Novel Approach to Negative Sampling with the Aim of Effectively Leveraging Entity Embeddings Obtained from the Preceding Batches. Within each iteration, $\mathbf{C}$ batches of previous representations are cached and utilized as negative samples for the subsequent iteration, contributing a total of $\mathbf{B} \bullet \mathbf{C}$ negative samples. Typically, only one or two pre-batches are employed [14]. We use two batches in our work.

Hard Negatives (HN) [18] Assumes Significance in Enhancing Contrastive Representation Learning, in Addition to Augmenting the Number of Contrastive Samples. Hard negatives are true negatives of high quality that are close to positive samples and hard to be distinguished, which contributes to training a good discriminative model. In line with [14], we set the case of head entity as the hard negative for tail entity prediction, which is motivated by the observation that textual encoding methods frequently assign a high score to the head entity primarily due to the presence of extensive textual overlap.

We combine the above three distinct types of negatives during training and denote their numbers as N(IN), N(PN), and N(HN), respectively. Consequently, we would have $N(\text{IN}) = \text{B} - 1$ in-batch negatives, $N(\text{PN}) = \text{B} \bullet \text{C}$ pre-batch negatives, and $N(\text{HN}) = 1$ hard negatives, resulting in a cumulative total of $\text{B}(\text{C} + 1)$ contrastive samples.

3.3 *Training Objective*

Most KGE models are trained using a margin-based loss that positive samples tend to have higher scores than negative ones. In order to effectively leverage current larger number of negatives of different types, we use InfoNCE loss with margin [28] as our optimization objective.

$$\text{loss} = -\log \frac{e^{(s(h,r,t)-\gamma)/\tau}}{e^{(s(h,r,t)-\gamma)/\tau} + \sum_{i=1}^{|\mathbb{N}|} e^{s(h,r,t_i')/\tau}} \tag{3}$$

$$t' \in \mathbb{N}(h, r) = \mathbb{N}_{\text{IN}} \cup \mathbb{N}_{\text{PN}} \cup \mathbb{N}_{\text{HN}} \tag{4}$$

The margin γ and the temperature τ are hyper-parameters. The additive margin $\gamma > 0$ benefits for assigning higher scores to correct triples (h, r, t). The temperature τ allows for adjusting the relative importance of negative samples. A smaller value of τ places more attention on hard negatives, yet causes a potential risk of over-fitting to label noise. $\mathbb{N}_{\text{IN}}$, $\mathbb{N}_{\text{PN}}$, and $\mathbb{N}_{\text{HN}}$ are sets of IN, PN, and HN, respectively. t' refers to the candidates for the tail entity to form negative samples. In general, the inclusion of a greater number of negative samples tends to result in more robust representations. Tests in earlier studies show that the temperature parameter has demonstrated a considerable impact on performance results [28], and the application of the margin parameter has consistently exhibited better performance compared to the standard InfoNCE loss, but the observed improvement is relatively modest in magnitude [14]. For fair comparison, we follow SimKGC [14] to set margin $\gamma = 0.02$ and temperature =0.05 .

4 Experiments

4.1 *Biomedical KGs*

In this work, we utilize two publicly accessible biomedical KGs, UMLS, and Hetionet, which are detailed in Table 2. **#E** and **#R** represent the numbers of entities and relations, respectively. UMLS [7] is a relatively small KG that encompasses medical semantic entities and their corresponding relationships. The majority

Table 2 Statistics of UMLS and Hetionet datasets

Dataset	#E	#R	#train	#dev	#test
UMLS	135	46	5216	652	661
Hetionet	47,031	24	1,800,157	225,019	225,021

of triples within UMLS describe widely recognized biological facts or medical commonsense. In contrast, Hetionet [3] is a large biomedical KG that combines knowledge from 29 publicly available resources. It is originally created for drug repurposing, by collecting connections among 11 types of biomedical entities such as compounds, diseases, and genes.

4.2 Comparison Methods

We compare our ensemble methods with multiple state-of-the-art KGC methods as follows: translational distance-based methods TransE [4] and its extensions TransH [5] and TransR [34], tensor factorization-based methods DistMult [6] and ComplEx [25], convolutional neural network-based methods ConvE [7] and ConvKB [35], graph neural network-based method R-GCN [36], pre-trained language model-based methods KG-BERT [8], StAR [9], PALT [10], and SimKGC [14].

4.3 Automatic Metrics

The evaluation of KGC methods primarily revolves around rank-based metrics including Hits@k (k $\in \{1, 3, 10\}$), Mean Rank (MR), and Mean Reciprocal Rank (MRR). For tail entity prediction of a given test triple (h, r, t), scores for all entities are ranked in a descending order to predict t based on h and r. Hits@k quantifies the proportion of ground-truth triples that appear within the top k highest-scored triples. Mean Rank (MR) calculates the arithmetic mean over ranks of all test triples, while Mean Reciprocal Rank (MRR) represents the arithmetic mean over the reciprocals of all ranks. And all these automatic evaluation metrics are reported under the filtered setting [4].

4.4 Experimental Settings

Our experiment is built upon LambdaKG [37], an open-source library for knowledge representation learning. The encoders in our methods are initialized with pre-trained checkpoints of biomedical PLMs from HuggingFace [38], as presented in Table 3. To determine the optimal learning rate, a grid search is conducted with a range of {1e-5, 3e-5, 5e-5}, and with 10 epochs for UMLS and 5 epochs for Hetionet.

Table 3 Names of register models used in our experiment

Model	Pre-trained checkpoints
BERT [12]	*Bert-base-uncased*
BioBERT [21]	*Dmis-lab/biobert-v1.1*
SciBERT [22]	*Allenai/scibert_scivocab_uncased*
PubMedBERT [23]	*Microsoft/BiomedNLP-PubMedBERT-base-uncased-abstract*
SapBERT [24]	*Cambridgeltl/SapBERT-from-PubMedBERT-fulltext*

The performance metrics for UMLS are sourced from the original papers of the respective methods being compared, while the implementation from the PyKEEN library [39] is employed to obtain results for Hetionet. All experiments are executed on a single NVIDIA RTX-3090 GPU.

5 Results and Analysis

5.1 Main Results

The experimental results are shown in Table 4. Since the UMLS dataset is much smaller than Hetionet, the evaluation metrics tend to exhibit larger numerical values for UMLS. We incorporate four distinct in-domain PLMs to form our ensemble methods to assess the effects of efficient contrastive learning and in-domain knowledge awareness. The suffix "CL" of the names of our methods stands for Contrastive Learning. A comprehensive analysis of the experimental results reveals that, the average performance of our ensemble methods is significantly better than the comparison methods.

Regarding graph embedding methods, traditional distance-based and semantic-matching-based methods demonstrate inferior performance compared to neural network-based methods. This disparity arises due to the capabilities of neural networks to express complex nonlinear projections and encode additional information. Consequently, neural network-based graph embedding methods can learn more robust representations and exhibit strong generalization capability. In contrast, DistMult performs poorly on both datasets because it restricts relation matrices to be diagonal. This limitation renders DistMult efficient only for symmetric relations, making it inadequate for general KGs. ComplEx extends DistMult by leveraging complex-valued embeddings, aiming to enhance the modeling of asymmetric relations. However, even with this extension, ComplEx does not perform well on the UMLS dataset.

Graph embedding methods exhibit significant competitiveness when compared to textual encoding methods. For instance, in the case of UMLS, TransH and TransR outperform PALT, KG-BERT, and StAR, as evidenced by their higher Hits@10 values. Similarly, on the Hetionet dataset, TransE, TransR, ConvE, and ConvKB

Table 4 Results of KGC task on UMLS and Hetionet datasets

	Method	UMLS		Hetionet			
		Hits@10↑	MR↓	Hit@1↑	Hits@3↑	Hits@10↑	MRR↑
	Graph embedding methods						
Translational distance	TransE	0.989	1.84	0.020	0.047	0.104	0.049
	TransH	0.995	1.80	0.017	0.040	0.096	0.043
	TransR	0.994	1.81	0.026	0.055	0.111	0.055
Semantic matching	DistMult	<u>0.846</u>	<u>5.52</u>	<u>0.006</u>	<u>0.016</u>	<u>0.054</u>	<u>0.025</u>
	ComplEx	0.967	2.59	–	–	–	–
Neural network	ConvE	0.990	1.51	–	–	–	–
	ConvKB	–	–	**0.030**	0.062	0.126	0.063
	R-GCN	–	–	0.019	0.061	**0.167**	0.064
	Textual encoding methods						
General PLMs	PALT	0.990	1.57	–	–	–	–
	KG-BERT	0.990	**1.47**	0.018	0.044	0.105	0.043
	StAR	0.991	1.49	–	–	–	–
	SimKGC	0.993	1.63	0.024	0.056	0.127	0.059
	Our ensemble methods						
In-domain PLMs	BioBERT-CL	**0.996**	**1.47**	0.026	0.059	0.133	0.063
	SciBERT-CL	0.992	1.53	**0.030**	**0.064**	0.144	**0.068**
	PubMedBERT-CL	0.993	1.52	0.027	0.060	0.134	0.063
	SapBERT-CL	0.993	1.51	0.026	0.061	0.135	0.064

surpass KG-BERT. However, SimKGC reverses such performance degradation as it leverages contrastive learning with a large negative sample size. Furthermore, the integration of in-domain PLMs contributes to a consistent improvement in the overall performance of our ensemble methods.

5.2 *Results Analysis*

We analyze the results in terms of the effects of efficient contrastive learning and the extent to which domain knowledge encapsulated into the model parameters of PLMs.

The Effects of Efficient Contrastive Learning In knowledge representation learning, negative sampling generates contrastive samples to facilitate training a discriminative model, which is a crucial step. It has been demonstrated that negative sampling significantly influences the quality of the learned representations [40].

Typically, embedding-based methods utilize many more negative samples than text-based methods, causing the performance degradation of the latter. Increasing the number of negative samples can lead to more robust representations [14], yet may require additional GPU memory and cause optimization difficulties [28]. By combining in-batch, pre-batch, and hard negatives and employing InfoNCE loss instead of conventional margin-based loss, both SimKGC and our ensemble methods perform well in the KGC task. The adoption of in-batch negatives advocates the reuse of entity embeddings, while pre-batch negatives are embeddings computed from an earlier version of model parameters. In contrast to the conventional embedding-based method RotatE [41], which employs approximately 64 negatives, and the typical text-based methods KG-BERT [8] and StAR [9], which utilize five negatives, the negative sampling strategy that combines three types of negatives used by our methods showcases simplicity and good training efficiency. Table 4 Results of KGC task on UMLS and Hetionet datasets.

The Amount of Domain Knowledge Stored in PLMS Recent studies have consistently demonstrated that the parameters of PLMs retain significant amounts of knowledge following self-supervised pre-training using extensive unlabeled text data. To address biomedical KGC, the use of in-domain BERT models offers distinct advantages compared to SimKGC, which employs a standard BERT model. The disparities among the four biomedical BERTs are reflected in their different pre-training processes. BioBERT exhibits the highest performance on the UMLS dataset, as it is first initialized from BERT of general domain, and subsequently undergoes further pre-training using PubMed abstracts and PMC articles. Given that the majority of triples in UMLS surround biomedical commonsense rather than rare professional terms, the world knowledge acquired from being initialized from BERT facilitates better understanding UMLS. The remaining three PLMs perform comparably and are solely focused on the biomedical domain, devoid of knowledge from the general domain.

SciBERT performs the best on the Hetionet dataset, followed by SapBERT and PubMedBERT. Hetionet is purely composed of biomedical terminology. SciBERT is pre-trained on PubMed Central full-text articles, and both PubMedBERT and SapBERT are only pre-trained on PubMed abstracts. Notably, PubMed abstracts contain approximately 4.5 billion words, whereas PubMed Central full-text articles contain around 13.5 billion words. The superior performance of SciBERT could be attributed to its larger pre-training corpus, which equips the model parameters with a greater volume of specialized knowledge. Furthermore, SapBERT exhibits better performance than PubMedBERT on both datasets due to its continual pre-training using UMLS synonyms, which especially benefits tasks related to entity-level information. High content overlap between pre-training corpora and downstream datasets would result in better model performance.

6 Conclusion

This paper introduces a novel approach that combines efficient contrastive learning with in-domain PLMs to enhance the performance of text-based methods in biomedical KGC. Our framework is built upon a dual-encoder architecture, where candidates for the encoder include BioBERT, SciBERT, PubMedBERT, and SapBERT. Additionally, we incorporate three distinct types of negative samples to enhance training efficiency without imposing substantial computational and memory burdens. Moreover, the compatibility between the existing knowledge within PLMs and the downstream tasks, as well as the alignment between the pre-training corpus and the task-specific datasets, significantly influence the ultimate performance of the proposed methods. For future work, we will expand our research by incorporating more biomedical KGs to further examine the generalizability and robustness of our ensemble methods in handling the task of biomedical KGC in addition to k-fold cross validation. To enhance the performance, we will explore alternative model architectures to represent the scoring function, and other types of negatives and other loss functions. Another direction will involve developing effective strategies to address false negatives during training due to the incompleteness of KGs.

Acknowledgments This research was supported by the Research Development Fund (RDFSP16) at Xi'an Jiaotong-Liverpool University (XJTLU), XJTLU Key Programme Special Fund (KSF-A-17), and the Suzhou Municipal Key Laboratory for Intelligent Virtual Engineering (SZS2022004).

References

1. Brown, A.S., Patel, C.J.: A standard database for drug repositioning. Sci. Data. **4**, 4 (2017)
2. Wishart, D.S., Knox, C.K., Guo, A., Cheng, D., Shrivastava, S., Tzur, D., Gautam, B., Hassanali, M.: DrugBank: a knowledgebase for drugs, drug actions and drug targets. Nucleic Acids Res. **36**, 901–906 (2007)
3. Himmelstein, D.S., Lizee, A., Hessler, C., Brueggeman, L., Chen, S., Hadley, D., Green, A.J., Khankhanian, P., Baranzini, S.E.: Systematic integration of biomedical knowledge prioritizes drugs for repurposing. elife. **6** (2017)
4. Bordes, A., Usunier, N., García-Durán, A., Weston, J., Yakhnenko, O.: Translating Embeddings for Modeling Multi-relational Data. NIPS (2013)
5. Wang, Z., Zhang, J., Feng, J., Chen, Z.: Knowledge graph embedding by translating on hyperplanes. In: AAAI Conference on Artificial Intelligence (2014)
6. Yang, B., Yih, W., He, X., Gao, J., & Deng, L.: Embedding Entities and Relations for Learning and Inference in Knowledge Bases. CoRR, abs/1412.6575 (2014)
7. Dettmers, T., Minervini, P., Stenetorp, P., Riedel, S.: Convolutional 2D knowledge graph embeddings. In: AAAI Conference on Artificial Intelligence (2017)
8. Yao, L., Mao, C., & Luo, Y.: KG-BERT: BERT for knowledge graph completion. ArXiv, abs/1909.03193 (2019)
9. Wang, B., Shen, T., Long, G., Zhou, T., & Chang, Y.: Structure-Augmented Text Representation Learning for Efficient Knowledge Graph Completion. Proceedings of the Web Conference 2021 (2020)

10. Shen, J., Wang, C., Yuan, Y., Han, J., Ji, H., Sen, K., Zhang, M., & Song, D..: PALT: Parameter-Lite Transfer of Language Models for Knowledge Graph Completion. Conference on Empirical Methods in Natural Language Processing (2022)
11. Vaswani, A., Shazeer, N.M., Parmar, N., Uszkoreit, J., Jones, L., Gomez, A.N., Kaiser, L., Polosukhin, I.: Attention Is All You Need. NIPS (2017)
12. Devlin, J., Chang, M., Lee, K., & Toutanova, K. (2019). BERT: Pre-training of Deep Bidirectional Transformers for Language Understanding. Conference of the North American Chapter of the Association for Computational Linguistics: Human Language Technologies
13. Liu, Y., Ott, M., Goyal, N., Du, J., Joshi, M., Chen, D., Levy, O., Lewis, M., Zettlemoyer, L., & Stoyanov, V.: RoBERTa: a robustly optimized BERT pretraining approach. ArXiv, abs/1907.11692 (2019)
14. Wang, L., Zhao, W., Wei, Z., Liu, J.: SimKGC: simple contrastive knowledge graph completion with pre-trained language models. In: Proceedings of the 60th Annual Meeting of the Association for Computational Linguistics (2022)
15. Oord, A.V., Li, Y., & Vinyals, O. (2018). Representation learning with contrastive predictive coding. ArXiv, abs/1807.03748
16. Karpukhin, V., Oğuz, B., Min, S., Lewis, P., Wu, L.Y., Edunov, S., Chen, D., Yih, W.: Dense passage retrieval for open-domain question answering. In: Conference on Empirical Methods in Natural Language Processing (2020)
17. Lee, J., Sung, M., Kang, J., & Chen, D. (2020). Learning Dense Representations of Phrases at Scale. Annual Meeting of the Association for Computational Linguistics
18. Gao, T., Yao, X., & Chen, D. (2021). SimCSE: Simple Contrastive Learning of Sentence Embeddings. Conference on Empirical Methods in Natural Language Processing
19. Petroni, F., Rocktäschel, T., Lewis, P., Bakhtin, A., Wu, Y., Miller, A.H., & Riedel, S. (2019). Language models as knowledge bases? Proceedings of the 2019 Conference on Empirical Methods in Natural Language Processing and the 9th International Joint Conference on Natural Language Processing
20. Roberts, A., Raffel, C., & Shazeer, N.M. (2020). How much knowledge can you pack into the parameters of a language model?. Conference on Empirical Methods in Natural Language Processing
21. Lee, J., Yoon, W., Kim, S., Kim, D., Kim, S., So, C.H., Kang, J.: BioBERT: a pre-trained biomedical language representation model for biomedical text mining. Bioinformatics. **36**, 1234–1240 (2019)
22. Beltagy, I., Lo, K., & Cohan, A. (2019). SciBERT: a pretrained language model for scientific text. Conference on Empirical Methods in Natural Language Processing
23. Gu, Y., Tinn, R., Cheng, H., Lucas, M.R., Usuyama, N., Liu, X., Naumann, T., Gao, J., Poon, H.: Domain-specific language model pretraining for biomedical natural language processing. ACM Trans. Comput. Healthcare. **3**, 1–23 (2020)
24. Liu, F., Shareghi, E., Meng, Z., Basaldella, M., & Collier, N. (2020). Self-Alignment Pretraining for Biomedical Entity Representations. North American Chapter of the Association for Computational Linguistics
25. Trouillon, T., Welbl, J., Riedel, S., Gaussier, E. & Bouchard, G. (2016). Complex embeddings for simple link prediction. Proceedings of the 33rd International Conference on International Conference on Machine Learning
26. Xie, R., Liu, Z., Jia, J., Luan, H., Sun, M.: Representation learning of knowledge graphs with entity descriptions. AAAI Conference on Artificial Intelligence. (2016)
27. Wang, X., Gao, T., Zhu, Z., Liu, Z., Li, J., Tang, J.: KEPLER: a unified model for knowledge embedding and pre-trained language representation. Trans. Assoc. Comput. Ling. **9**, 176–194 (2019)
28. Chen, T., Kornblith, S., Norouzi, M., Hinton, G.E.: A simple framework for contrastive learning of visual representations. Proceedings of the 37th International Conference on Machine Learning. **119**(149), 1597–1607 (2020)
29. Mikolov, T., Sutskever, I., Chen, K., Corrado, G.S., Dean, J.: Distributed representations of words and phrases and their compositionality. Adv. Neural Inform. Process. Syst., 26 (2013)

30. He, K., Fan, H., Wu, Y., Xie, S., Girshick, R.B.: Momentum contrast for unsupervised visual representation learning. Conference on Computer Vision and Pattern Recognition, pp. 9726–9735 (2019)
31. Reimers, N., & Gurevych, I. (2019). Sentence-BERT: sentence embeddings using siamese BERT-networks. Conference on Empirical Methods in Natural Language Processing
32. Zan, D., Chen, B., Yang, D., Lin, Z., Kim, M., Guan, B., Wang, Y., Chen, W., & Lou, J. (2022). CERT: continual pre-training on sketches for library-oriented code generation. International Joint Conference on Artificial Intelligence
33. Qian, J., Li, G., Atkinson, K., Yue, Y.: Understanding negative sampling in Knowlege graph embedding. Int. J. Artif. Intell. Appli. **12**, 71–81 (2021)
34. Lin, Y., Liu, Z., Sun, M., Liu, Y., Zhu, X.: Learning entity and relation embeddings for knowledge graph completion. AAAI Conference on Artificial Intelligence. (2015)
35. Nguyen, D.Q., Nguyen, D.Q., Nguyen, T.D., Phung, D.Q.: A convolutional neural network-based model for knowledge base completion and its application to search personalization. Semantic Web. **10**, 947–960 (2019)
36. Schlichtkrull, M., Kipf, T., Bloem, P., Berg, R.V., Titov, I., & Welling, M. (2017). Modeling relational data with graph convolutional networks. Extended Semantic Web Conference
37. Xie, X., Li, Z., Wang, X., Zhu, Y., Zhang, N., Zhang, J., Cheng, S., Tian, B., Deng, S., Xiong, F., Chen, H.: LambdaKG: a library for pre-trained language model-based knowledge graph embeddings. ArXiv, abs/2210.00305. (2022)
38. Wolf, T., Debut, L., Sanh, V., Chaumond, J., Delangue, C., Moi, A., Cistac, P., Rault, T., Louf, R., Funtowicz, M., & Brew, J. (2019). HuggingFace's transformers: state-of-the-art natural language processing. ArXiv, abs/1910.03771
39. Ali, M., Berrendorf, M., Hoyt, C.T., Vermue, L., Sharifzadeh, S., Tresp, V., & Lehmann, J. (2020). PyKEEN 1.0: a python library for training and evaluating knowledge graph embeddings. ArXiv, abs/2007.14175
40. Kotnis, B., & Nastase, V. (2017). Analysis of the impact of negative sampling on link prediction in knowledge graphs. ArXiv, abs/1708.06816
41. Sun, Z., Deng, Z., Nie, J., & Tang, J. (2018). RotatE: knowledge graph embedding by relational rotation in complex space. ArXiv, abs/1902.10197

Oracle-Inspired RFID-Based Indoor Positioning Architecture for Sharing Assets

Zuoyang Qu, Wei Hui, Weiqiao Zhu, Lun Su, Deqiang Kong, and Weimeng Wang

1 Introduction

RFID indoor positioning combined with blockchain technology has diverse applications across various industries, including supply chain management, asset tracking, smart buildings, and retail. These applications share several commonalities, including real-time tracking, data integrity, automation, efficiency, and security [1].

In the context of the Ethereum whitepaper [2], an oracle machine is a method used to incorporate external data into smart contracts for data verification and blockchain operations. Oracles act as intermediaries between off-chain and on-chain sources, facilitating the retrieval of data from external sources and passing it to smart contracts for processing. The whitepaper emphasizes the importance of security, trustworthiness, and robustness in oracles, along with the need for a reliable mechanism to validate and transmit external data.

2 Related Technology

From previous studies [3], it can be found that among various indoor positioning technologies, positioning using RFID has the advantages of small sensor size and low power consumption, making it the preferred technology for locating valuable assets such as securities. To establish a decentralized RFID indoor positioning system, it is crucial to transmit off-chain location information to the blockchain in a synchronous or asynchronous manner. The location data obtained from the

Z. Qu (✉) · W. Hui · W. Zhu · L. Su · D. Kong · W. Wang
Institute of Computing Technology, China Academy of Railway Sciences, Beijing, China
e-mail: quzuoyang@rails.cn

X. Jia et al. (eds.), *2nd International Conference on Big Data, IoT, and Cloud Computing*, EAI/Springer Innovations in Communication and Computing,
https://doi.org/10.1007/978-3-031-93570-1_5

RFID sensor array can be encapsulated within an API interface, enabling the provision of external services [4]. By utilizing a blockchain platform equipped with smart contracts, the dissemination of location information on the blockchain can be achieved by incorporating additional fields into a user's account transaction request [5]. However, since on-chain smart contracts cannot actively retrieve off-chain data, the presence of an oracle mechanism is essential in this environment. The oracle mechanism facilitates interaction between smart contracts and external data sources, enabling data access and execution triggering [6]· One approach involves deploying an off-chain service responsible for retrieving data from various sources and transmitting it to the on-chain contract through a script. This script can operate at specific time intervals or respond to the state of on-chain nodes.

The implementation of a centralized oracle for data transmission to the blockchain presents a risk of a single point of failure [7]. When smart contracts rely on data from a centralized service, overall security is diminished as the contract becomes dependent on a single source. Centralized oracles are vulnerable to network failures or downtime, which can disrupt business operations and hinder timely data access for users. Furthermore, centralization introduces the possibility of the oracle supplying malicious data for its own advantage, especially in decentralized finance (DeFi) protocols or non-fungible token (NFT) projects [8].

To address these issues, decentralized blockchain oracles have been designed. Decentralized oracles eliminate reliance on a single source of truth, enhancing the authenticity of information provided to smart contracts [9]. Unlike centralized oracles, decentralized oracles rely on multiple external sources and aim for trustlessness. They employ mechanisms such as the ShellingCoin mechanism, where independent sources report data without coordination. However, decentralized oracles face challenges such as collusion, signaling, and bribery, which must be carefully considered and addressed to ensure effectiveness and integrity [10]. For a more comprehensive comparison between Centralized Oracle and Decentralized Oracles, as shown in Table 1.

Commonly known blockchain oracle projects include Chainlink, Band Protocol, Tellor, Augur, and Provable [11]. Based on the requirements stated in the Ethereum whitepaper, new oracle projects can also be established. In our research focused on trustworthy RFID positioning data on the blockchain [12], this research utilized an external adapter based on Chainlink to achieve the goal. Chainlink is a network of decentralized oracles that provide external data for smart contracts on the blockchain [9]. The digital asset used for payment on the Chainlink network is LINK tokens. A Chainlink network consists of Chainlink nodes with registered job specifications, which perform job execution coordinated by on-chain oracle contracts. LINK tokens serve as incentives for Chainlink node operators to serve customers through Chainlink customer smart contracts.

An oracle contract has functionality to allow the architecture that involves off-chain and on-chain to work. The first part controls which Chainlink clients also have invitations to fulfill work requests, another role would be to be a gateway of requested client nodes to the Chainlink network. Chainlink nodes are responsible for executing the requested jobs [9].

Table 1 Comparing centralized oracle with decentralized oracles

Index	Domain	Centralized oracle	Decentralized oracles
1	Data source	Single or multiple	Single or multiple
2	Exchange fees	Higher fees	Less fees
3	Mode	Single Oracle node	Multiple Oracle node
4	Cost Struct	Gans fees	Gans fees
5	Cost impact factor	Code&Preference	Code
6	Authentication	Third party	Multisig&Certificate
7	Efficiency	Higher speed	Lower speed
8	Risk	Trusted validation Single point of failure	Data contamination Sybil attack Mirroring attack
9	Currency	USD	Token
10	Hardware platform	TEEs	TEEs
11	Security	More prone to hacks and data leaks	Less prone to hacks and data leaks
12	Stages	Startup	Startup
13	Scalability	Lower scalability	Higher scalability

Compare the other off-chain indoor position technology, through RFID to garget fixing position can take more advantages than the other [13–17]. RFID is cost-effective and at the same time it can provide highly accurate location data, often within a few centimeters. By the way, RFID systems can easily be scaled up to track a large number of items and no line of sight required. Those advantages are concluded with the verdict, becoming an attractive solution for the asset chain to provide indoor positioning data.

RFID-based indoor positioning is indeed a cost-effective solution, primarily due to the low cost of passive RFID tags [18]. These tags can be easily deployed as reference tags for positioning purposes. By interrogating the tags and recording their responses, we can uniquely determine the position of the RFID reader or the mobile platform carrying the reader. This RFID-based positioning scheme has the potential to provide higher accuracy compared to other wireless indoor positioning schemes.

RFID-based indoor positioning systems typically comprise antennas, tags, readers, and positioning algorithms [19]. Common algorithms used in RFID-based indoor positioning include Time of Arrival (TOA), Time Difference of Arrival (TDOA), Angle of Arrival (AOA), and Received Signal Strength Indication (RSSI). In this research, the relative position of the target is determined based on the prior information of the RFID tag array topology, building upon the concept of LANDMARC [20]. The exact positions of the RFID tags are known in advance and stored in a database. When a reader or another tag is in proximity, its position relative to the known tags can be determined. This is achieved by measuring the signal strength or time of arrival of the signals from the surrounding tags and comparing these measurements with the stored information about the tag positions.

Even in complex indoor environments, this method can provide accurate positioning information based on the relative distances to multiple known tags.

In this study, an external adapter suitable for RFID array positioning hardware-framework with corresponding on-chain contracts are designed to monitor the asset's location for P2P user which cannot use centralized location services due to crisis of confidence.

3 Positioning System Design

There is such a situation in business activities where assets frequently circulate among multiple peers, the relationships between each peer are equal, based on mutual trust and supervision. When an asset is created by its creator and its flow scope is specified (i.e., where it can circulate among which entities), the asset becomes collectively owned by all entities within the flow scope and is subject to supervision from all stakeholders [21]. There are many such environment s, such as centralized sharing of Teslas. In these environment s, each entity has its own independent asset positioning network. Due to the flexible compatibility and equal relationships, it is difficult for a centralized positioning solution to flexibly assist each entity organization in transparently and fairly accessing real-time location information of the shared valuable assets. Therefore, the decentralized RFID-based positioning technology designed in this research can effectively meet the requirements of this environment [22].

Before discussing the design, some concepts must be declared. Which are limited to the scope of this research.

Asset: A physical object that can be processed or produced and has value for multiple groups.

Peers: An organization that operates independently in terms of group behavior, without dependency or subordination relationships. It has the autonomy to make decisions regarding internal affairs and possesses its own independent on-chain account.

Peer Union: Peer Union refers to a collective formed by peer entities that jointly own one or more assets.

Positioning nodes: Logical nodes of an RFID indoor positioning system owned by each peer entity.

The Peer Union means a decentralized on-chain community comprised of peer entities that Agree on the value of an asset. Within this alliance, each member deploys arrays of RFID positioning sensors in their respective spatial domains. These sensor arrays facilitate the capture of the position of a specific tag within the member's spatial domain. The acquired position information can then be exchanged on the blockchain.

A shared Oracle contract has been utilized to connect the sensor arrays and positioning nodes to the same blockchain. The positioning node cluster includes a

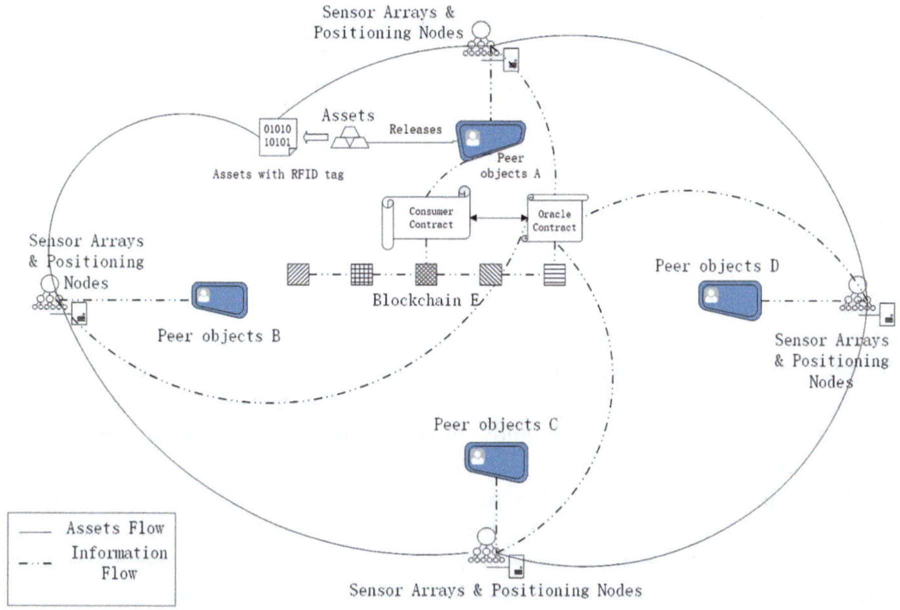

Fig. 1 Overall system design diagram

Chainlink Node, which becomes part of the distributed oracle network. As depicted in Fig. 1, Peer Objects A to D form a Peer Union and establish a connection to the blockchain E through an Oracle Contract.

Peer Object A generates a unique hash value for the asset before releasing it, which is based on the asset's attributes and timestamps. This hash value is then written onto an RFID tag that is attached to the physical asset. The RFID tag serves as a crucial identifier for the asset on the blockchain. Additionally, Peer Object A deploys a consumer contract on the blockchain. When the asset, tagged with the RFID, moves within the Peer Union, the consumer contract receives location requests from member users' accounts. These requests include the asset's hash value as a parameter. Subsequently, the contract forwards the hash value as an event to the Oracle Contract.

In a classical setting, Peer Objects A, B, C, and D establish a Peer Union and deploy an oracle contract on Blockchain E. This oracle contract has been registered as the Chainlink nodes of Peer Objects A, B, C, and D. Peer Object A produces a valuable asset and designates it as Assets I after affixing an RFID tag to the asset. Concurrently, a consumer contract is deployed on Blockchain E, and the contract's address is shared within the Peer Union. The consumer contract's role is to receive user-initiated requests and query the Oracle Contract for the position of Assets I.

The asset is subsequently circulated among Peer Objects A, B, C, and D, while being bound to the RFID tag and adhering to specific circulation rules. The asset remains within the spatial domains of the peer objects, and during this period, the positioning nodes continuously capture and record the position information of the tag signals within their detection range. These positioning nodes then supply the most recent position data of the queried asset in response to the contract's query request.

3.1 Adapter Design

Adapters typically provide data to smart contracts in the form of web APIs. In the case of RFID-based indoor positioning, the adapters use RFID antennas to gather tag location information. This information is then aggregated and exposed as a service through an API. The communication distance between the tags and antennas limits RFID indoor positioning. In this study, multiple antennas are used to aggregate the positioning information.

When tags communicate with readers, transmission information such as signal strength is generated. By placing a fixed number of tags in the environment, the signal characteristics of these tags can be continuously obtained. When a new tag appears in the environment, it will exhibit signal characteristics similar to the closest fixed tag. By comparing the similarity between the fixed tags and the new tag, several closest positions to the new tag can be identified. Finally, the position of the new tag can be estimated [23]. Alternatively, by collecting prior information about the tags' presence in the environment, a set of prior information can be formed. During actual positioning, the estimated position of the target can be obtained by comparing the similarity between the target information and the prior information. This method avoids the need to estimate distances, which can be affected by factors such as multipath interference.

Based on the above, a positioning method based on scene analysis has been token. Figure 2 below illustrates the scene concept.

This design for an RFID localization adapter focuses on the systematic deployment of reference tags in the localization environment. Both the reference tags and target tags take the same parameters, but the system has prior knowledge of the reference tag positions. By comparing the single character parameters of tags of the same type, it becomes possible to determine the approximate area where the target tag is located. This information, when combined with other algorithms, enables more accurate and precise localization results.

In the localization scenario, there exist n reader antennas and m reference tags. The RSSI vector of a reference tag, measured by n antennas at coordinates (xi, yi), is denoted as Ri = (R1, R2, R3, ..., Rn), where i ranges from 1 to m. Similarly, the RSSI vector of the target tag is represented as T = (T1, T2, T3, ..., Tn), and its coordinates are (x, y). The similarity between the target tag and the i-th reference tag can be expressed as follows:

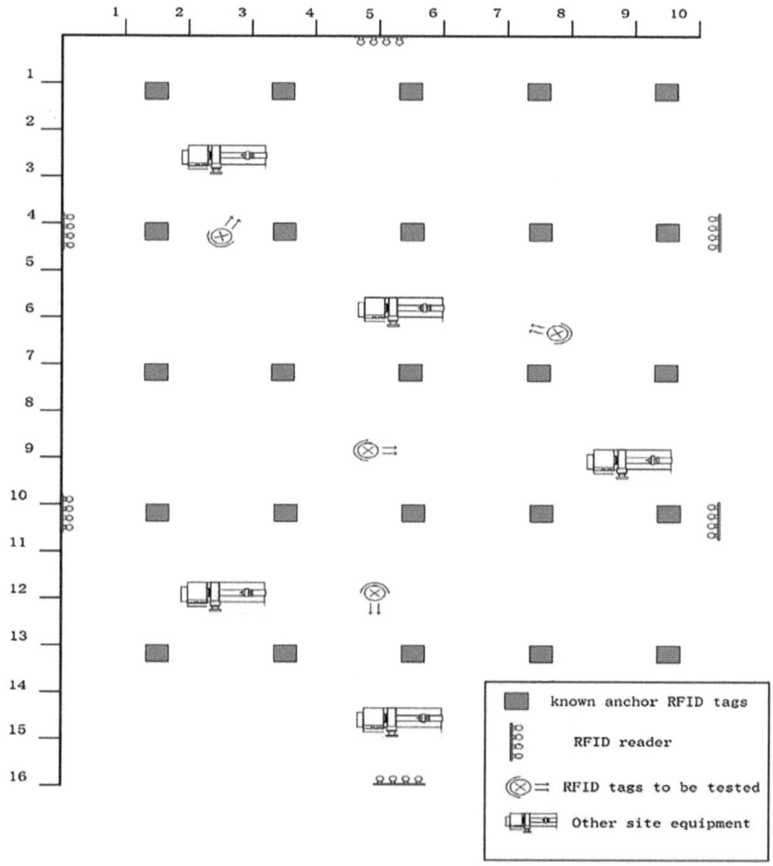

Fig. 2 RFID indoor positioning adapter design

$$E_i = \sqrt{\sum_{j=1}^{n} (T_j - R_i)^2} \tag{1}$$

The RSSI vector value between the target tag and the nearest reference tag indicates proximity, as closer reference tags yield smaller Euclidean distances. When utilizing K nearest reference tags, the normalization weight coefficient of the Kth nearest reference tag can be expressed as follows:

$$\omega_k = \frac{\frac{1}{E_k^2}}{\sum_{k=1}^{K} \frac{1}{E_k^2}} \tag{2}$$

From Eq. (2), it can be observed that the nearest reference tag obtains the highest weight coefficient. By using the weight ratio, the target coordinates can be estimated as:

$$(x, y) = \sum_{k=1}^{K} \omega_k (x_k, y_k) \tag{3}$$

After processing, the coordinates will be transformed into a physically meaningful and interpretable location information.

An RFID indoor positioning adapter array is constructed within the spatial range of an equivalent object by deploying reference tags, reader antennas, and other devices. This array collects positional data from multiple sources to obtain accurate asset location information. Although this data aggregation may appear centralized, each equivalent object can only access its own internal positional information, ensuring decentralized operation. This approach avoids transmitting all positional information to a single location, maintaining a decentralized design for the entire adapter array.

The deployment of numerous reader antennas and reference tags, along with IoT device control signals, multipath coordination, and data aggregation operations, requires substantial computational and network resources. These tasks cannot be efficiently performed using on-chain computing or serverless computing, as they involve mapping the internal spatial positions of equivalent objects. Therefore, it is recommended that each equivalent object deploys dedicated Oracle Nodes to address these requirements.

3.2 *Positioning Node Design*

A positioning node is a resource management node for an equivalent object, comprising Chainlink nodes, adapters, and sensor arrays. Due to the varying internal spatial sizes of peer objects, a significant deployment of RFID antennas, readers, and Chainlink nodes may be required. The obtained positioning coordinates through the adapter indicate the relative positions within a peer object and lack practical significance in the external space. The positions of multiple sensors need to be aggregated for computation. Hence, a comprehensive positioning node framework is essential.

The positioning node depicted in Fig. 3 consists of three logical architecture layers. The top layer is the Web Service layer, responsible for resource management. It facilitates management functions for peer objects to the nodes, including the general functionalities of the MIS system. Node Controller accesses the Web Service layer through an HTTPS app, providing a graphical interface. The Controller manages metadata, status information, contract addresses, user accounts, and automated scripts for request handling. The Register enables registration, discovery

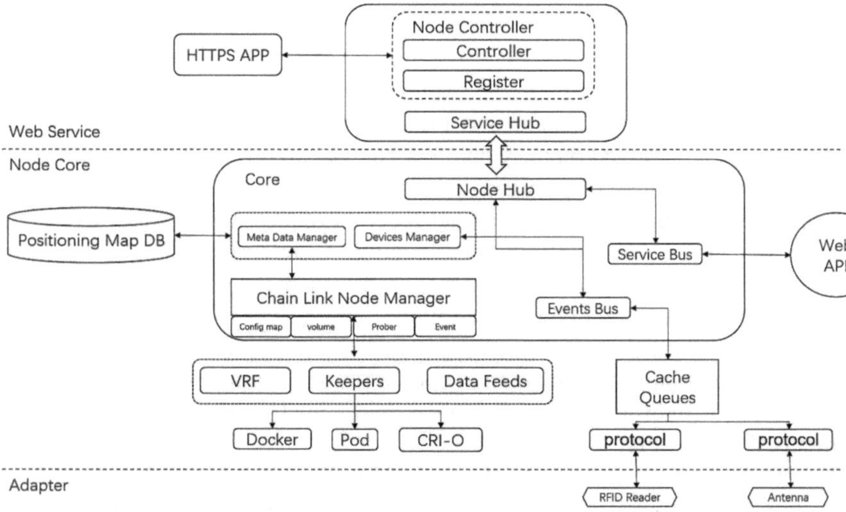

Fig. 3 Localization node architecture design

of Node Cores, and information routing. The Service Hub and Node Hub establish communication tunnels using various protocols to ensure seamless information distribution between the controller and Register. The upstream data flow synchronizes Node Core's status information, while the downstream data flow disseminates configuration instructions, user requests, and other data. The Node Core layer offers a distributed and highly available runtime environment for Chainlink Nodes, along with a stable Adapter interface. This design enhances the stability of the distributed Chainlink oracle. The Chainlink Node Manager oversees a distributed cluster of Chainlink Nodes and configures resources based on descriptive files, serving as the foundation for running Chainlink Node instances. Docker, Pod, and CRI-O enable the instantiation of Chainlink Nodes, ensuring network stability. The RFID reader writes tag location information to cache queues using industrial control protocols, triggering positioning messages. It obtains relative coordinates and passes them to the Metadata Manager. The Positioning Map DB maps the relative coordinates to physical location information, which is then uploaded to the Web Service layer of the corresponding equivalent object through the Node Hub. The location service is provided via external adapters in the form of service buses. The Devices Manager not only handles the underlying devices of Chainlink Nodes but also responds to information like RFID reader frequency, power, and phase. In cases where there are limitations in the number of RFID reader interfaces and the site scale, multiple RFID readers may be deployed. Hence, Node Core ensures the management of multiple RFID readers and the ordering of messages.

3.3 Contract Design

Off-chain node devices utilize Oracle contracts and user contracts to facilitate the upload of data to the blockchain and provide on-chain data services [24]. The Oracle contract is deployed on the Chainlink Nodes of the Node Core, enabling the transaction-driven upload of positioning information from the RFID adapter to the blockchain. Meanwhile, the user contract receives requests from users within the physical alliance and retrieves the corresponding asset's positioning information from the Oracle contract using a transaction-driven messaging mechanism. The design of these two contract types is as follows:

The Oracle contract's primary class, named Oracle, is derived from Ownable and employs SafeMath for fundamental operations. It incorporates three interfaces: ChainlinkRequestInterface, OracleInterface, and LinkTokenInterface. Their class diagram is shown as Fig. 4.

The Oracle contract is collectively owned by the members of the peer union, which allows for flexible and dynamic management of these members. The Ownable contract facilitates this ownership structure. The implementation of the ChainlinkRequestInterface enables the handling of data request interfaces for Chainlink. It allows for the sending of request messages to external adapters wrapped through the service bus. The OracleInterface is utilized to implement the callback functionality for data responses. This allows the Oracle contract to receive the aggregated RFID tag location information through the callback function defined in the OracleInterface. Lastly, the LinkTokenInterface implements standard interfaces for interacting with the LINK token, providing methods for transfer, query, and authorization. Additionally, the Oracle class itself incorporates functions for dynamic balancing of multiple data sources and verifying the validity of data source addresses.

The user contract makes requests to the Oracle for asset location information based on transactions in the user's account. As depicted in Fig. 5, the user contract interacts with the Oracle contract through the PositionSale class, which facilitates transaction-driven location queries. When the user account initiates a transaction request, it obtains asset location information from this contract. The PositionSale class implements the Aggregatorv3interface interface to enable data interaction with

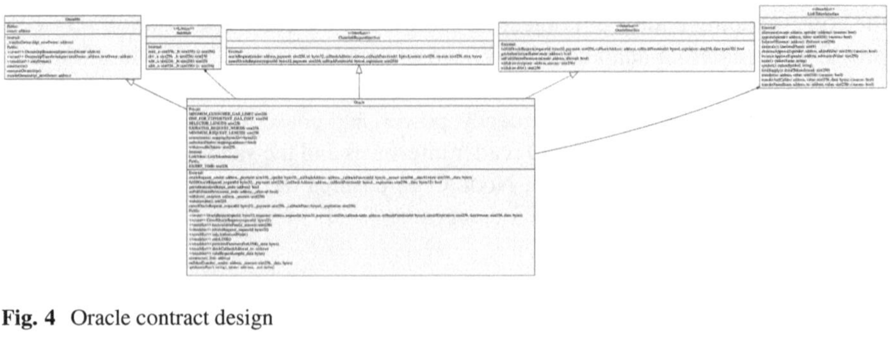

Fig. 4 Oracle contract design

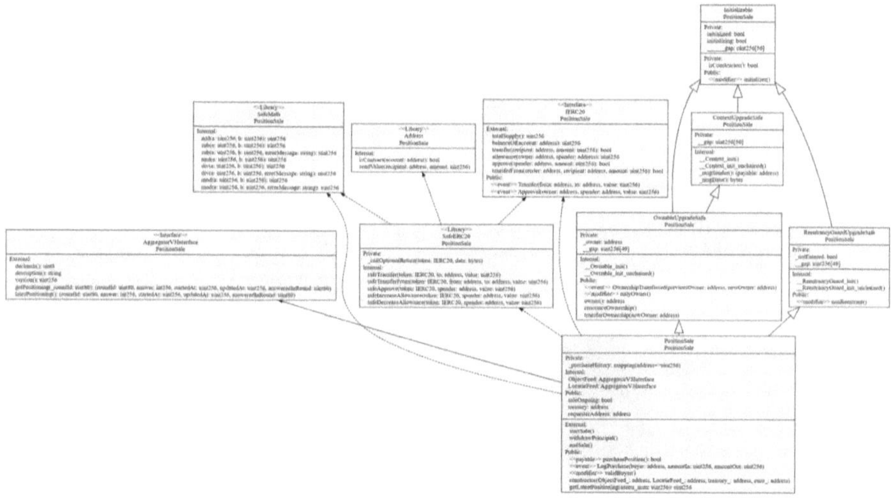

Fig. 5 Customer contract design

the Oracle contract and retrieve the most up-to-date asset location information. It utilizes IEC20, Safe ERC20, and other implementations for handling transactions. PositionSale acts as an aggregation proxy for location information, allowing peer union members to initiate transactions with it to acquire the current asset location information.

4 Case Study

In this scenario, a document serves as an asset, and its SHA-1 hash value, along with a timestamp, is used as an RFID tag ID ("97B902A3D7A28130A18E54F487BEC18 471CA5D52"). This RFID tag is affixed to physical paper documents, as depicted in Fig. 6. A set of antennas is deployed within the indoor space of the peer, with a known anchor RFID tag securely attached to the desktop. A positioning node is created by combining a radio receiver board and a personal computer. The personal computer is also connected to an RFID reader service, which interfaces with the antenna array. Additionally, this personal computer serves as a chain link node, forming the hardware architecture for this particular case.

The RSSI of each RFID tag on the tables is initially measured individually. This measurement allows for the establishment of a correspondence between known anchor tags and the actual physical locations in the environment. The collected information is stored in the Positioning Map Database. Subsequently, an asset with an RFID tag is introduced into the experimental area and moves within it. As a result, an RSSI matrix is generated and stored in the adapter, representing the signal strengths observed at different locations. This matrix can be visualized as a

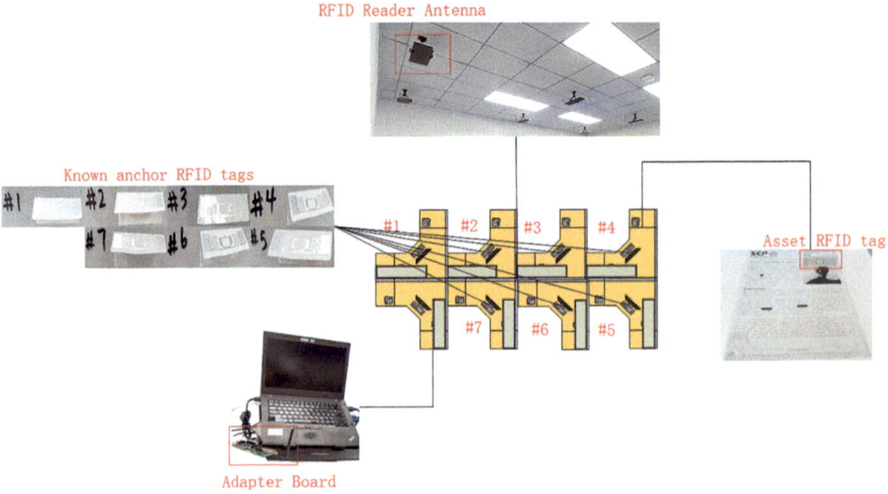

Fig. 6 Case study environment

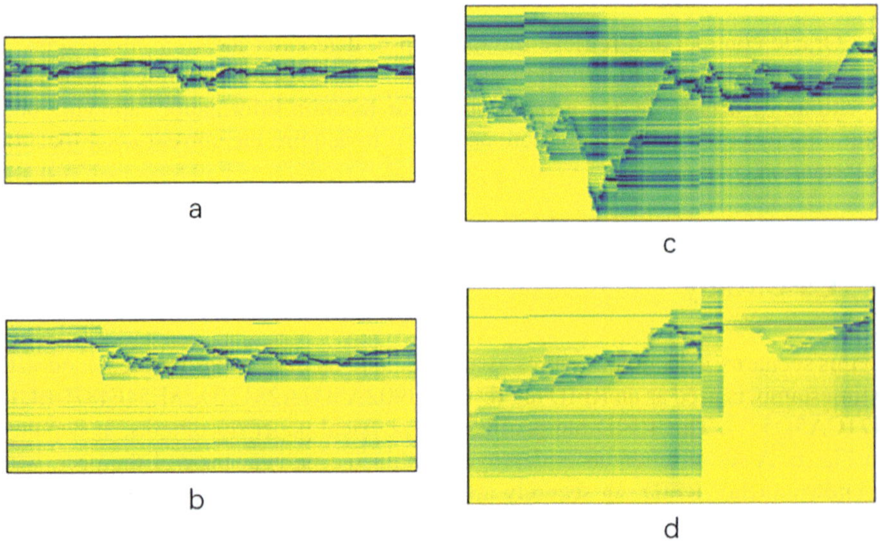

Fig. 7 Experiment area asset RFID tag RSSI

heatmap; the location information of the four files is depicted in Fig. 7. The RSSI data is then correlated with the information in the Positioning Map DB, enabling the derivation of the asset's physical location within the Node core. This information is subsequently uploaded to the blockchain network through the oracle. Figure 8 illustrates the capability to obtain positioning information on the test blockchain network.

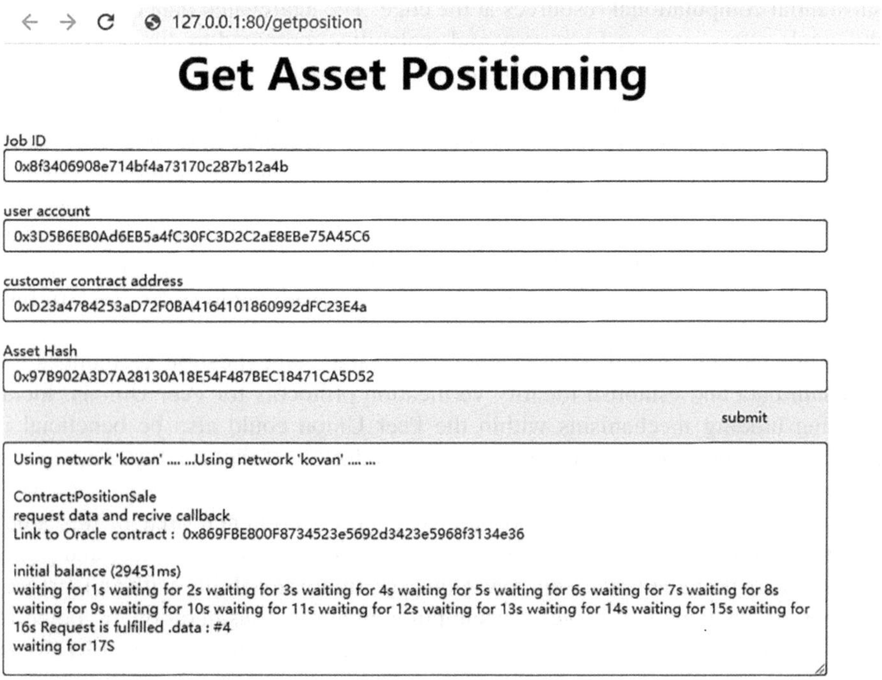

Fig. 8 Querying assert location and return the result

5 Conclusions and Discussion

This research proposes a method for flexible asset tracking using Chainlink and an RFID tag array. The method leverages an alliance formed by users querying a specific asset, where each member shares their asset location information. This allows for complete asset lifecycle tracking, with decentralized and easily scalable information sharing.

The system consists of three layers. The first layer includes RFID antennas, readers, and signal strength decoding boards situated at the edge. In the experimental setup, software-defined radio devices can integrate reader and signal strength decoding functionalities. The second layer comprises an upper computer that connects and consolidates the edge data, retaining the mapping between known regions, RSSI values, and physical locations. The upper computer also acts as a Chainlink node for deploying an Oracle. The third layer encompasses the on-chain contract component, where consumer contracts on the blockchain receive user requests and retrieve aggregated asset location information from the Oracle contract.

The use of RFID and blockchain for sharing location information sets this research apart from other blockchain-based supply chain initiatives. Continuous location updates, especially for indoor positioning of passive devices, require substantial computational resources at the edge. The aggregated data obtained from the oracle offers improved accuracy and scalability compared to direct positioning achieved through tag antenna combinations.

As elaborated in the literature review mentioned earlier, the use of RFID-based indoor positioning technology can address the underlying positioning technology needs for thin and lightweight objects in low-power indoor scenarios. Another key element of this paper, which combines RFID indoor positioning technology with blockchain, is the adoption of the oracle mechanism during the data-on-chain process. Compared with other blockchain-based RFID indoor positioning solutions, the introduction of the oracle mechanism can effectively address the issue of data on-chain.

Future studies should explore encryption and authentication techniques for on-chain data and establish identity verification protocols for Peer Unions. Incorporating bidding mechanisms within the Peer Union could also be beneficial in handling unexpected node disconnections or nodes of inferior quality. These areas present opportunities for further investigation.

Furthermore, it is important to note that the experimental area in this study covered approximately 40 sq. m., where the demand for wireless devices and power requirements remained relatively low. However, future scalability with an increased number of devices and power consumption requires consideration of potential interference from other electromagnetic devices nearby, which can affect RF signal recognition. To mitigate noise, remote filtering and signal processing techniques can be employed.

Acknowledgments In this paper, the research was sponsored by Science and technology research and development program of China National Railway Group Co., LTD.(N2022S012).

References

1. Paul, T., Mondal, S., Islam, N., Rakshit, S.: The impact of blockchain technology on the tea supply chain and its sustainable performance. Technol. Forecast. Soc. Chang. **173**, 121163 (2021)
2. Buterin, V., et al.: Ethereum: A next-generation smart contract and decentralized application platform. https://github.com/ethereum/wiki/wiki (2014)
3. Hayward, S., Lopik, K.V., Hinde, C.J., West, A.A.: A survey of indoor location technologies, techniques and applications in industry. Internet Things. **20**, 100608 (2022)
4. Wang, C.Y., Xie, L., Zhao, Y.-C., Zhang, D.-Q., Ye, B.-L., Lu, S.-L.: Survey on RFID-based battery-less sensing. Journal of Software. **33**(1), 27 (2022) (in chinese)
5. Zhang HaiQiang, D., Rong, A.S.: Intellectual property management model based on ethereum smart contract. Sci. Technol. Manag. Res. **15**, 164–169 (2021) (in Chinese)
6. Adler, J., Berryhill, R., Veneris, A., Poulos, Z., Veira, N., Kastania, A.: Astraea: a decentralized blockchain oracle. In: Proceedings of the 2018 IEEE International Conference on Internet of Things (iThings) and IEEE Green Computing and Communications (GreenCom) and IEEE Cyber, Physical and Social Computing (CPSCom) and IEEE Smart Data (SmartData), pp. 1145–1152 (2018)

7. Woo, S., Song, J., Park, S.: A distributed Oracle using Intel SGX for Blockchain-based IoT applications. Sensors. **20**(9) (2020)
8. Kaleem, M., Shi, W.: Demystifying Pythia: a survey of ChainLink Oracles usage on ethereum. (2021). https://doi.org/10.48550/arXiv.2101.06781
9. Lorenz Breidenbach.: Chainlink 2.0: Next Steps in the Evolution of Decentralized Oracle Networks https://research.chain.link/whitepaper-v2.pdf (2021)
10. Merre R (2020). Safety beyond the grave through Chainlink and NGRAVE. DOI:https://doi.org/10.13140/RG.2.2.21360.79369
11. Al-Breiki, H., Rehman, M.H.U., Salah, K., Svetinovic, D.: Trustworthy blockchain oracles: review, comparison, and open research challenges. IEEE Access. **8**, 85675–85685 (2020)
12. YuHao, H., Jie, Z., ZeXin, L., Hong, Z., YueZhou, D., ChunYue, H.: Design and implementation of trusted digital asset deposit system based on blockchain. Netw. Security Technol. Appli. **4**, 14–17 (2021) (in Chinese)
13. Luo, J., Zhang, Z., Chang, L., et al.: Reliable and cooperative target tracking based on WSN and WiFi in indoor wireless networks. IEEE Access. **99**, 1–1 (2018)
14. Hernandez, N., Corrales, H., Parra, I., et al.: WiFi-based urban localisation using CNNs. In: 2019 IEEE Intelligent Transportation Systems Conference—ITSC (2019)
15. Kalbandhe, A.A., Patil, S.C.: Indoor positioning system using bluetooth low energy. In: 2016 International Conference on Computing, Analytics and Security Trends (CAST) (2016)
16. Konings, D., Budel, A., Alam, F., et al.: Entity tracking within a Zigbee based smart home. In: 2016 23rd International Conference on Mechatronics and Machine Vision in Practice (M2VIP), pp. 1–6 (2016)
17. He, S., Chan, S.: Wi-Fi fingerprint-based indoor positioning: recent advances and comparisons. IEEE Commun. Survey. Tutor. **18**(1), 466–490 (2017)
18. Fang, Q., Huang, J., Wang, P., Zhang, Z., Li, S.: Indoor fixed assets localization method based on RFID and machine learning. J. Guangxi Univ. Sci. Technol. **33**(03), 43–45 (2022) (in Chinese)
19. Abdulrahman, A., Abdulmalik, A.S., Mansour, A., et al.: Ultra wideband indoor positioning technologies: analysis and recent advances. Sensors. **16**(5), 1–36 (2016)
20. Yeh, C.H., Su, S.F.: Enhance LANDMARC from the fundamentals. In: 2013 International Conference on Advanced Robotics and Intelligent Systems, pp. 23–27 (2013)
21. Biswas, K., Muthukkumarasamy, V., Lum, W.: Blockchain based wine supply chain traceability system. In: Proceedings of the Future Technologies Conference (FTC), pp. 56–62 (2017)
22. Feng, X.: Design and Implementation of Government Procurement System Based on Ethereum Smart Contract. Jilin University (2021) (in Chinese)
23. Liu, X., Wen, M., Qin, G., et al.: LANDMARC with improved k-nearest algorithm for RFID location system. In: 2016 2nd IEEE International Conference on Computer and Communications (ICCC), pp. 2569–2572 (2016)
24. ChainLink.: Chainlink developer documentation. https://docs.chain.link (2023)

Research on the Application of Blockchain Technology in Trusted Federated Learning

Dilu Zhang, Qing Zhang, and Zhongchen Miao

1 Introduction

With the explosion of applications for AI large models, the importance of privacy protection and data security has become increasingly prominent. Clarkson Law Firm stated that OpenAI secretly scraped a large amount of unauthorized personal information from the Internet, affecting millions of users. In August 2023, The New York Times updated its terms of service to prohibit the use of its content for training AI models. Data misuse can harm individual rights, but excessive data protection can hinder the release of data value. Therefore, finding legal and compliant solutions to address data fragmentation and silos is a significant challenge in the era of large-scale AI models.

Federated learning [1], as a new paradigm of distributed machine learning, has become a key technological approach to balance data usage and privacy security. Its ability to operate on invisible data enables participants to conduct collaborative model training while protecting their data security and user privacy, thereby exploring data value and breaking down data silos. However, in the coordination process among multiple participants, the credibility of data sources, training processes, and training results becomes an important dimension for evaluating the feasibility of engineering applications of federated learning. Blockchain technology, as a multi-centered trust system, naturally aligns with the characteristics of federated learning, providing new solutions for achieving trusted federated learning. This article will first introduce the characteristics of federated learning and the challenges it faces, and then explore the characteristics of blockchain technology and its application research in trusted federated learning.

D. Zhang (✉) · Q. Zhang · Z. Miao
Shanghai Financial Futures Information Technology Co., Ltd., Shanghai, China
e-mail: zhangdl@cffex.com.cn

X. Jia et al. (eds.), *2nd International Conference on Big Data, IoT, and Cloud Computing*, EAI/Springer Innovations in Communication and Computing,
https://doi.org/10.1007/978-3-031-93570-1_6

2 Characteristics and Challenges of Federated Learning

2.1 *Overview of Federated Learning*

Federated learning is an innovative collaboration mechanism that enables unified model training on multi-party data while ensuring data privacy and security [2]. During the training process, each client's raw data is stored locally and is not exchanged or directly transmitted. Instead, encrypted parameters are exchanged through instantaneous aggregation to achieve iterative model updates.

Training large models involves diverse samples and multidimensional data, which are commonly fragmented and distributed across multiple parties. Traditional multi-party data processing models often involve simple data transactions, where one party directly hands over user data to another party for model training. However, such data transactions become challenging to sustain under existing legal protections and regulations. The value of federated learning lies in its ability to establish a disruptive collaborative process, constructing a compliant way for data value circulation in the era of large-scale models, with "data availability without visibility" as its core principle [3].

2.2 *Challenges of Trusted Federated Learning*

At the application level, federated learning relies on cryptographic algorithms such as secure multi-party computation [4], homomorphic encryption [5] to achieve encrypted operations on model parameters among multiple parties. However, to achieve fully trusted federated learning among multiple participants, it is necessary to consider not only the security of cryptographic algorithms but also the potential malicious behavior of participants. This article mainly focuses on the challenges faced by trusted federated learning in terms of data sources, training processes, and training results.

Trusted Data Sources In federated learning, each client's raw data is stored locally. In the absence of a trusted third-party coordinating entity for data assessment and auditing in the collaborative environment, data may face compliance risks and malicious poisoning risks [6]. Compliance risks refer to the difference from traditional centralized modeling, where participants only need to ensure the compliance of their own data. In federated learning, all parties need to ensure that their data has a complete authorization chain. Otherwise, risks from one party's data may spread as global risks. Malicious poisoning risks refer to the possibility of malicious participants introducing invalid or corrupted data samples into the federated dataset, thereby degrading the model's performance or prediction accuracy, rendering it incapable of correctly handling real-world tasks. Specific attack methods include using redundant data to reduce model generalization ability, randomizing labels deviating from the true model, embedding backdoors to access private data, etc.

Trusted Training Process Federated learning training relies on the aggregation and updating of model parameters from all parties, involving processes such as encrypted sample alignment and encrypted gradient exchange. Therefore, malicious attackers may pollute the global model by injecting malicious parameters or gradients [7]. Specifically, malicious attackers alter intermediate gradients or send incorrect information during the model update process, maximizing the deviation between the global model's direction and the direction before the attack to achieve the desired attack effect. Since each party receives securely aggregated and encrypted results, they cannot know how other participating parties generate their local model updates. As a result, traditional detection methods generally cannot effectively identify if other clients' parameters are abnormal by directly detecting the aggregated parameters. Apart from malicious model poisoning attacks, participants may also attempt to infer the original data through reconstruction attacks and other means.

Trusted Training Results Federated learning produces joint outcomes among multiple parties. However, different organizations may have varying levels of business development, varying data quality, and different feature distributions, leading to a lack of reasonable and trusted mechanisms for evaluation and benefit allocation of the model outcomes. If the cost of trust is too high or the coefficient of cooperation friction is too significant, multi-party collaboration cannot be established, making it difficult to achieve economies of scale. On the other hand, due to the involvement of hardware, software, network security, and many implementation details in federated learning, there may be extreme cases where participants engage in illegal or non-compliant behavior, such as misuse of federated models. Therefore, to enhance trust from participating parties and regulatory authorities towards training results, effective supervision and post-audit evaluation mechanisms need to be provided.

3 Application of Blockchain in Trusted Federated Learning

3.1 Overview of Blockchain

Blockchain technology utilizes a specific chain structure and cryptographic algorithms to create a decentralized trust mechanism. Simply put, a blockchain is formed by sequentially connecting blocks, where each block is like a database, and writing data to a block is recording data in the database. The consensus mechanism in a blockchain network ensures that attackers cannot simultaneously modify multiple block information, which means that data becomes difficult to maliciously tamper with after being written into the blockchain network.

3.2 Blockchain-Based Trusted Federated Learning

Blockchain's tamper resistance and transparency can be applied to create a secure and transparent multi-party trust system, where different participants can collaborate while maintaining trust in the integrity of data and transactions. Both federated learning and blockchain utilize cryptography, distribution, and other technologies, sharing similarities while having their own unique features. By combining federated learning with blockchain, it is possible to achieve verifiability, traceability, auditability, and evaluation in the workflow, laying a foundation for achieving end-to-end trustworthiness.

Data Traceability and Behavior Recording The capability of identity and behavior management based on blockchain can be applied to strengthen the security of federated learning platforms and participants, further enhancing trust among data collaborators. For the issue of multi-party data authorization in federated learning, distributed digital identities [8] implemented on the blockchain can standardize and authorize data assets. Technologies such as verifiable credentials, incorporating digital signatures and zero-knowledge proofs, can ensure safer and more trustworthy claims about data sources, providing infrastructure support for fine-grained permission control. To mitigate the risk of data poisoning, technical means and consensus mechanisms can be used to perform cross-validation of critical data involved in collaborative computations on the blockchain, recording the critical states of the data. Additionally, data sources from participating parties can be stored on the chain for evidence purposes, enabling traceability and verification through hash value validation.

Information Processing and Workflow Execution Blockchain-based smart contracts enable automated workflow execution and detection, effectively preventing and identifying abnormal behavior of participants during the training process. On the one hand, the business processes of federated learning can be expressed on the blockchain through smart contracts, standardizing participant behavior and ensuring the integrity and reliability of the computation process through pre-auditing, public transparency, and automatic execution algorithms. On the other hand, for model poisoning attacks during the training process, blockchain can record the behaviors of participating parties during data sharing. By introducing an evaluation and automatic monitoring system based on smart contract behavior quality, it can promote secure and trustworthy training processes. Specifically, monitoring the changes in predicted values of the global model through the blockchain can determine which client's aggregated value caused those changes. Smart contracts can identify and discard suspicious data by detecting anomalies in the locally uploaded gradients from clients. Combining automatic detection results with the consensus process enables behavioral management of participants, increasing the cost of malicious actors engaging in misconduct.

Incentive Allocation and Contribution Evaluation A well-designed incentive mechanism will increase the motivation of all parties involved in the federated learning system. For the trained models, smart contracts can update training costs based on metrics such as the number of data uses and computational consumption. For subsequent inference and prediction processes of the model, smart contracts can precisely limit the value of data usage to specific algorithms and usage counts, achieving controllable and measurable purposes. Participating parties can also store fingerprints of each data request and invocation on the blockchain, ensuring that query records and data are genuinely traceable. When regulatory authorities have compliance concerns about the model, participating parties can access model usage records on the blockchain to ensure the authenticity and trustworthiness of the data used in the process. Additionally, the usage records of the model can serve as proof of contribution among participating parties, facilitating the distribution of benefits among organizations.

3.3 Algorithm Flow Analysis

The federated learning process based on blockchain is illustrated in Algorithm 1 and consists of the following main steps:

Algorithm 1 Federated Learning with Blockchain

```
Initialize federated learning parameters, model, and
network structure.
Contract_FL ← InitializeFederatedLearningSmartContract()
Contract_Contribution ← InitializeContributionSmartContract()
Contract_Anomaly ← InitializeAnomalyDetectionSmartContract()
T, η, λ, e ← InitializeHyperparameters()
for each participant do
    Identity ← GenerateDigitalIdentity()
    StoreDataProvenanceAndFingerprintOnChain(Data_k, Identity)
end for
while not ConvergenceCondition() do
    for each participant do
        LocalModel ← LocalTrainingPhase(Dataset_i)
        Enc_Gradients, Enc_Labels ← HomomorphicEnerypt(LocalModel)
        UploadToBlockchain(Enc_Gradients, Enc_Labels, Contract_FL)
    end for
    Val_Gradients ← ValidateGradientsOnBlockchain(Contract_Anomaly)
    Agg_Gradients ← AggregateGradients(Val_Gradients,Contract_FL)
    DistributeAggregatedGradientsToParticipants(Agg_Gradients)
    NewGlobalModel ← GenerateUpdatedGlobalModel()
    for each participant do
        ReceiveGlobalModelFromBlockchain(Contract_FL)
        PerformNextRoundOfLocalTraining()
    end for
end while
FinalModel ← FinalizeGlobalModel()
EvaluateContributionAndRewards(Contract_Contribution)
PerformOutcomePredictionAndValueAllocation()
RegulatoryReviewAndComplianceAssessment()
```

1. Participating parties negotiate the various processes of federated learning and initialize parameters, including federated learning smart contracts, contribution evaluation smart contracts, anomaly detection smart contracts, as well as iteration rounds (T), learning rate (η), regularization parameter (λ), homomorphic encryption key (e), etc.
2. Each participating party traces the data sources and fingerprints on the blockchain using digital identities to ensure compliance and integrity of the source data. After the consensus process, they register the smart contracts and initiate the federated learning process.
3. In each training round, each client trains a local model based on local data and the federated learning smart contract. They send encrypted gradient parameters and predicted labels to the blockchain network through homomorphic encryption. Consensus nodes use the anomaly detection smart contract to identify and discard suspicious data from the clients' uploaded gradients. The aggregated gradients are redistributed to each client for the next round of model training until the federated learning model converges.
4. Participants obtain the converged federated learning model, and consensus nodes automatically evaluate the contributions of each client during the model training based on the contribution evaluation algorithm. The goal of the contribution evaluation algorithm is to achieve Pareto-optimal resource allocation, where each participant's incentive depends on their level of contribution. Contribution evaluation metrics can include accuracy rates of predicted labels compared to actual labels or evaluation based on chip capacitance parameters, CPU cycles, and data size.
5. Based on the predicted data and the federated learning model, participants collaborate to make result predictions and perform subsequent value distribution. Regulatory authorities can conduct post-audits and compliance assessments based on the process records on the blockchain.

3.4 Experiment

We implemented the algorithm based on the FATE [9] platform, compared and evaluated the dataset from a simulated real-world vertically federated learning application using both local training and federated learning training approaches. The data interaction process during training was implemented through smart contract algorithms. To evaluate the accuracy, efficiency, and security of the federated learning process, this experiment designed classification accuracy indicators and training time indicators.

As shown in Fig. 1, this federated model achieved comparable results to the local training, but incurring efficiency loss as detailed in Fig. 2. After three epochs, the model gradually converged, as evidenced by the classification accuracy. The blockchain-based federated learning model achieved comparable results to the local model. Comparing the training time reveals that the blockchain-based federated

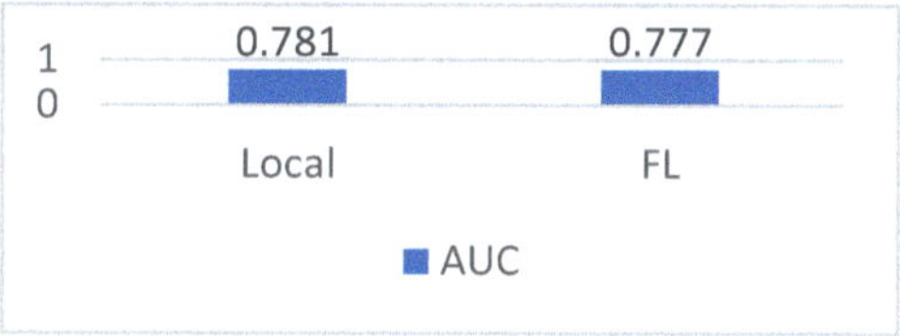

Fig. 1 Model prediction performance comparison

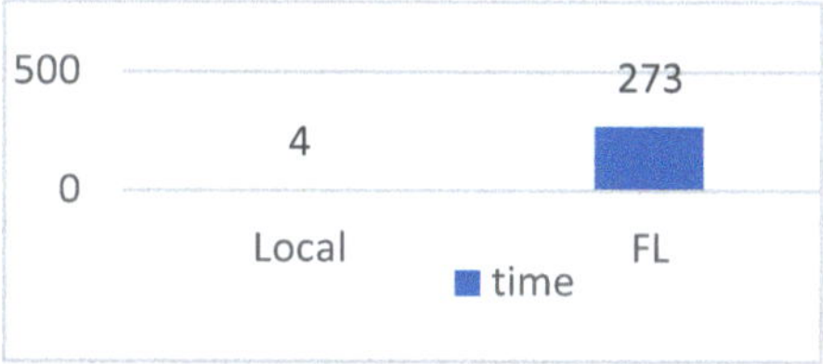

Fig. 2 Training efficiency comparison

learning model incurs some efficiency loss, but it remains within the acceptable range for industrial applications. From the perspective of algorithm security, packet capture analysis shows that the training process of federated learning is conducted with encrypted data transmission. Data reveals that the user processing and model usage information is all recorded and tamper-proof, achieving traceable and trusted federated learning.

4 Summary

Blockchain-based trusted federated learning offers significant advantages in data traceability, process compliance, and incentive allocation. However, it is important to address the potential risks associated with blockchain technology itself as a new source of risk in the data flow process. Sufficient maturity and security in consensus and data communication are necessary for it to become the infrastructure of the data asset market. Regulatory institutions also need to clarify the responsibilities, obligations, and risk mitigation mechanisms for automated federated learning processes based on smart contracts at the legal and regulatory levels and undertake rigorous evaluations and experiments in terms of institutional frameworks and regulatory approaches.

References

1. Konen, J., Mcmahan, H.B., Yu, F.X., et al.: Federated Learning: Strategies for Improving Communication Efficiency [J] (2016). https://doi.org/10.48550/arXiv.1610.05492
2. Yang, Q., Liu, Y., Chen, T., et al.: Federated machine learning: concept and applications [J]. ACM Trans. Intelli. Syst. Technol. **10**(2), 1–19 (2019). https://doi.org/10.1145/3298981

3. Fernandez, J.D., Potenciano, S., Lee, C.M., et al.: Privacy-preserving federated learning for residential short-term load forecasting [J]. Appl. Energy. (2022)
4. Canetti, R., Lindell, Y., Ostrovsky, R., et al.: Universally Composable Two-Party and Multi-Party Secure Computation [J]. In: Proc. STOC'02, pp. 494–503 (2002). https://doi.org/10.1145/509907.509980
5. Fan, J., Vercauteren, F.: Somewhat practical fully homomorphic encryption [J]. Iacr Cryptol Eprint Archive. (2012)
6. Zhao S, Ma X, Zheng X, et al. Clean-Label Backdoor Attacks on Video Recognition Models [J]. 2020
7. Jere, M.S., Farnan, T., Koushanfar, F.: A taxonomy of attacks on federated learning. IEEE Security Privacy. **19**(2), 20–28 (2020)
8. Dunphy P, Garratt L, Petitcolas F. Decentralizing digital identity: open challenges for distributed ledgers [C] 2018 IEEE European Symposium on Security and Privacy Workshops (EuroS&PW). IEEE, 2018: 75–78
9. Liu, Y., Fan, T., Chen, T., et al.: Fate: an industrial grade platform for collaborative learning with data protection [J]. J. Machine Learn. Res. **22**(1), 10320–10325 (2021)

A Low-Cost Government-to-Government Data Sharing Architecture Based on Blockchain

Yuexin Zuo, Yifan Wu, Cheng Li, Hua Zhang, Chen Ye, and Tingcong Ye

1 Introduction

In recent years, the transformation of government functions has become a research hotspot. The shift from governments being "managers" to "service providers" and the enhancement of government service efficiency are the directions in which governments worldwide are striving. To achieve this, a plethora of information systems has been developed to improve the efficiency of various government departments. However, more and more government affairs involve with different departments, and these government affair transactions require interdepartmental data requests and state updates. Therefore, the challenge of establishing data sharing and interaction among those mutually independent information systems belonging to different departments has become a widespread issue. The development of blockchain technology has provided a new perspective for government data sharing. This paper constructs a low-cost and easily deployable Government-to-Government (G2G) [1] data sharing architecture based on blockchain technology. It not only ensures data security but also enhances data interaction efficiency, while it keeps the cost of applying blockchain technology to existing systems low. To this end, this paper makes the following contributions:

Y. Zuo
HDU-ITMO Joint Institute, Hangzhou Dianzi University, Hangzhou, China

Y. Wu (✉) · C. Li · H. Zhang · C. Ye
School of Computer Science and Technology, Hangzhou Dianzi University, Hangzhou, China
e-mail: yfwu@hdu.edu.cn

T. Ye
School of Electronics and Information Engineering, Hangzhou Dianzi University, Hangzhou, China

X. Jia et al. (eds.), *2nd International Conference on Big Data, IoT, and Cloud Computing*, EAI/Springer Innovations in Communication and Computing,
https://doi.org/10.1007/978-3-031-93570-1_7

1. By combining Hierarchical Identity-Based Encryption (HIBE) [2] algorithms with Role-Based Access Control (RBAC) [3], data security is ensured during the data sharing process, enabling role-based fine-grained data access.
2. Transforming system operation entities from government employees into government roles facilitates the isolation of data sharing system from the original information systems, which only requires minor modification to the original systems, thereby reducing the whole development and deployment costs.
3. Leveraging the traceable characteristics of blockchain technology enables traceability in government data access requests, addressing the problem of determining responsibility when data breaches occur.
4. We have designed and validated smart contracts based on this architecture, providing a feasible solution for government data sharing.

2 Related Work

Yiwei Zhang [4] analyzed the research on blockchain technology in government data sharing from both a technical and a managerial perspective. Junfang Zeng [5] compared the advantages of blockchain technology over data centers and proposed the theoretical structure of a blockchain bus, achieving logical unity and physical dispersion in architecture. However, [4, 5] have shortcomings in that they only provide general conceptions which lack technical details and experimental validation. Noe Elisa [6] introduced a framework for a decentralized e-government peer-to-peer (P2P) system using blockchain technology to ensure information security and privacy while increasing mutual trust among public departments. Chunhui Piao [7] proposed a Service-On-Chain (SOC) method. The SOC method effectively identifies data retrieval needs from different departments, efficiently shares government data, and ensures both data content trustworthiness and data ownership control. However, [6, 7] do not consider the application of access control and data encryption algorithms on the blockchain. Furthermore, the above-mentioned studies are all based on developing entirely new systems integrating with blockchain technology to replace existing information systems. In practice, abandoning these existing systems and developing a new blockchain system is often economically unfeasible for many local governments.

Several researchers have explored the integration of the HIBE and RBAC with blockchain. Zhiguo Wan [8] proposed HIBEChain and applied it in the field of the Internet of Things (IoT), ensuring high scalability and accountability mechanisms for IoT devices. Jason Paul Cruz [9] introduced a role-based access control (RBAC-SC) platform based on smart contracts, utilizing Ethereum's smart contract technology for cross-organizational role utilization. However, [8] primarily focuses on the IoT domain and lacks research in permission control [9]. uses RBAC and provides a contract, but it lacks encryption algorithms resulting in insufficient security. Junfang Zeng [3] also mentioned using RBAC to convert object permissions into role permissions to reduce system complexity but did not provide detailed design information regarding the integration of RBAC.

3 G2G Data Sharing Architecture

3.1 Requirements of Government Data Sharing

To address the limitations of the above research, we propose a data sharing architecture targeted at resource-constrained local governments. Due to constraints related to funding, talent, and equipment, these local governments can only afford to adopt new technologies within a manageable cost range to enhance their governance efficiency. Thereby, we put forward the following requirements for G2G data sharing:

1. Easy Deployment: The architecture should adhere to the principles of Blockchain as a Service (BaaS) [10] and be easy to deploy, with the goal of achieving data sharing with minimal impact on existing systems.
2. Mutual Isolation: The blockchain data sharing system and the local governments' original systems should be mutually isolated to reduce mutual interference between the two systems.
3. Security and Reliability: Security and reliability of government data exchange should be ensured through permission control and encryption algorithms.
4. Transparency and Openness: Basic functionalities for data traceability and data transparency must be provided to avoid issues of unclear responsibilities.

3.2 Solution

To meet the above requirements, the architecture we have designed provides the following solutions:

1. We adopt the BaaS approach and achieve government data sharing among departments by developing standardized data interfaces on top of existing information systems.
2. We transform system operation entities from government employees into government roles, thereby achieving isolation between the data sharing system and the original system.
3. We ensure data security from two dimensions: data encryption and access control, achieved through the combination of the HIBE algorithm and RBAC model.
4. We guarantee transparency and openness in the data sharing process by storing data request proofs and state updates of government affairs in the blockchain.

3.3 Architecture

The blockchain architecture for Government-to-Government (G2G) data sharing that we propose is a further improvement upon the general architecture introduced in [11]. The schematic diagram of the proposed architecture is shown in Fig. 1.

Initialization We assume that there exists a responsible government department in charge of deploying and maintaining the blockchain, while other departments will participate in and use the blockchain. To initialize the government data sharing architecture, the first step is for all the participating departments to jointly establish the standards of the shared data, including the standards for government affair transactions. They also need to grant corresponding data permissions to

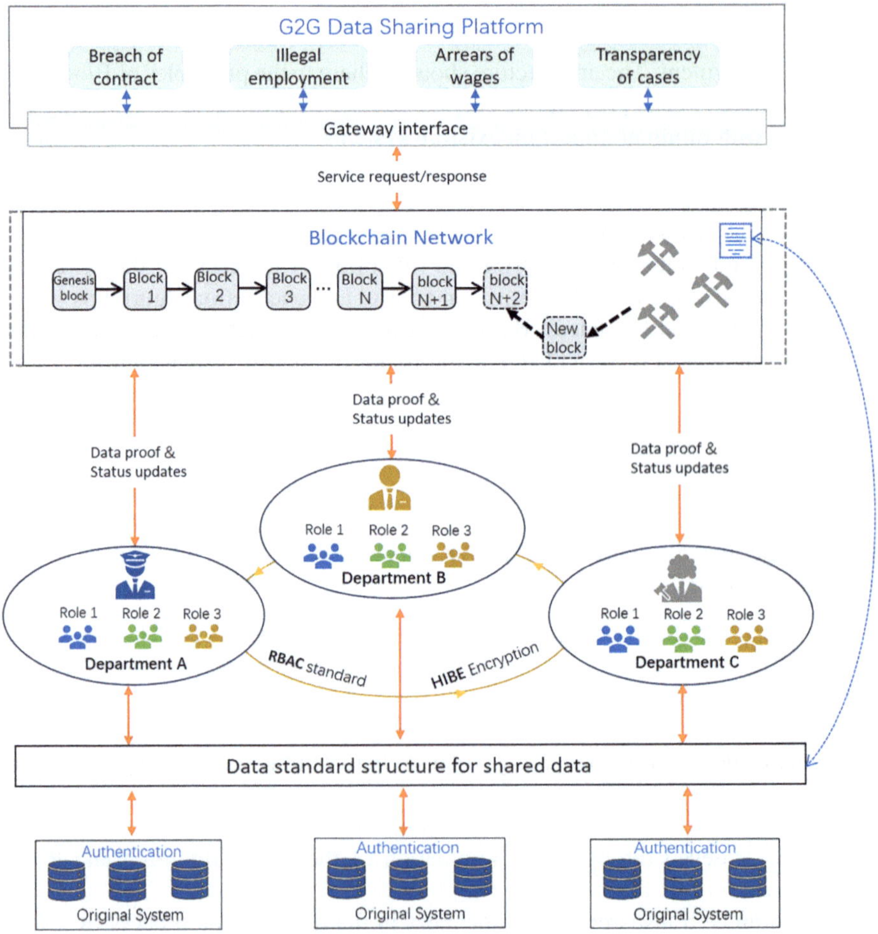

Fig. 1 G2G data sharing architecture based on blockchain

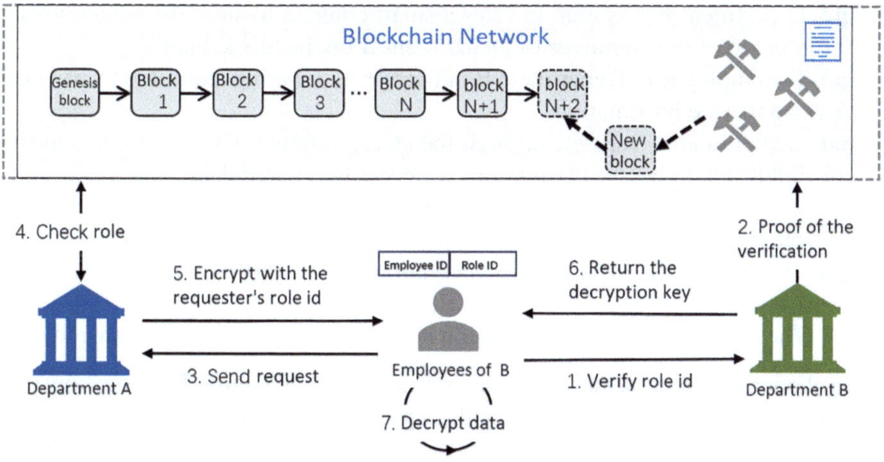

Fig. 2 Government data flow process

various government roles and formulate the (role, data permission) mappings. These mappings will be recorded on the blockchain and made public to all departments. However, the (employee, role) mappings for the relationship between a specific employee and a government role are kept within the departments, ensuring that personnel changes within the departments do not affect the data sharing system. This use of RBAC for data access control provides a standardized and modular solution for government data sharing. It also isolates the original system from the data sharing system, reducing the cost of integrating blockchain into the existing information systems.

Data Sharing This stage is the core of the entire architecture. It's worth noting that balancing data sharing efficiency with data security has always been a research focus in government data sharing. Some complex encryption algorithms like Ciphertext-Policy Attribute-Based Encryption (CP-ABE) have been utilized to protect data security [12]. These algorithms provide fine-grained data access but often require extensive modifications to the original databases, which many local governments cannot afford due to prohibitively high development costs [13]. The key focus of this paper is how to ensure fine-grained data access and enhance data security while reducing the cost of integrating blockchain systems. To achieve this, we avoid encrypting the original government data but, instead, bind roles to identity information in HIBE. By appropriately dividing the data permissions possessed by roles, we achieve fine-grained data access and data security, providing a low-cost solution for G2G data sharing. Figure 2 is a simple example of data request and response, divided into the following steps:

Step1. An employee from Department B authenticates itself as a Department B role through the internal authentication system of Department B.

Step2. Department B's system invokes a smart contract to store the authentication information of the employee of Department B on the blockchain.
Step3. The employee of Department B initiates a data sharing request to Department A using its role information.
Step4. Department A verifies, through the smart contract, whether the requesting role holds the necessary permissions to access the shared data.
Step5. Upon successful verification, Department A encrypts the shared data using the role ID and returns it to the initiating employee of Department B.
Step6. Department B provides the decryption key to the initiating employee of Department B.
Step7. The employee uses the decryption key to decrypt the encrypted shared data.

In the above process, the blockchain primarily serves three key functions: (1) data access control, (2) traceability, and (3) government affairs state update.

4 Smart Contract Design and Verification

4.1 Smart Contract Design

In this section, we introduce the smart contracts designed for the proposed architecture. The smart contracts are coded in the Solidity programming language. The source code of the contracts is publicly available on GitHub.

DataAccessControl Contract This smart contract provides permission-related information for authorization and verification of shared data.

Initialization: Writes the initial (role, data permission) mappings into the smart contract.
CheckPermission(Field) Public Returns (Bool): Verifies whether the caller has access to the specified field, returns the verification result, and triggers an event to log the verification request.
Updatepermission(Field, Role, HasPermission): Updates the (role, data permission) mapping. Only the authorization center can call this function.

RoleIdentityRecord Contract This smart contract is used to generate or query the proof of ownership of the roles an employee has acquired.

addRecord(Userid, Roleid, RoleAccount): Takes the employee ID (userId), role ID (roleId), and the blockchain account of the role (roleAccount) as input to create an identity authorization record and stores it in the blockchain, used to record the role identity and corresponding role account obtained by each employee. Only called by the contract owner.
getRecordsByRoleId (RoleId) Returns (Record[ ]): Takes an identity (roleId) as input to query when and by which employee this identity has been used.

AffairsFactory Contract This smart contract is used to create government affair transaction subcontracts and record the history of state changes and actions of transactions.

createAffairs (Name, Desc): This function is called through the factory contract to take the government affairs name (name) and associated description (desc) as input to create the government subcontract and record the subcontract address.

updateState (NewState, Remark): This function takes newState and remark as input to update the transaction states and save the operation record for traceability.

GetTraceInfo(): This function is used to query the history information of the government affairs. It returns a history of operations (traceData) to trace the state of the affairs.

4.2 Smart Contract Verification and Gas Cost

To verify the feasibility of the aforementioned contracts, we compiled and deployed them on the Polygon (Ethereum layer 2 network) and Conflux. We analyzed the costs of deploying smart contracts on both chains and calculated the average cost. During our analysis, we chose the medium execution rate of the current network to set the gas value, using 1MATIC $\approx$ 0.62 USD and 1CFX $\approx$ 0.15 USD in October 2023, as presented in Table 1. Our observations revealed that the average cost of deploying smart contracts on the blockchain was around 0.0005 USD. However, this cost is a one-time expense for setting up and initializing the system, while the costs of executing other functions are relatively low. Nevertheless, it can be argued that these costs will be significantly lower compared to buying, setting up, and maintaining a new database for sharing.

5 Conclusion

In the field of e-government, improving the efficiency of interdepartmental data sharing and affairs handling, enhancing the transparency of government affairs, and safeguarding the security and privacy of government data have always been important research directions. The emergence of blockchain technology has provided new solutions for e-government. However, the pressing issue remains how to utilize blockchain technology at a low cost. In this paper, we propose an architecture that provides a secure, traceable, and easily deployable platform for G2G data sharing. By combining HIBE and RBAC, we transform the system's operation entities from government employees into government roles, offering a standardized and modular solution while ensuring security. We have designed and made public a group of prototype smart contracts for this architecture and tested them on the Polygon and Conflux Network, demonstrating the feasibility of the proposed method. In the

Table 1 Gas cost for deployment of the smart contracts

	Polygon Gas Used	Conflux Gas Used	Polygon Tx Fee	Conflux Tx Fee	Polygon Tx USD	Conflux TX USD	Average USD Cost
DataAccessControl	550,743	612,347	0.001377	0.000612	0.000716	0.000067	0.000555
RoleIdentityProof	394,001	472,315	0.000985	0.000472	0.000052	0.000052	
AffairsFactory	1,385,414	1,675,432	0.003464	0.001675	0.000184	0.000184	

future, the proposed method can be extended to data sharing solutions between the government and citizens (G2C) or between the government and businesses (G2B).

References

1. Elisa, N., Yang, L., Li, H., Chao, F., & Naik, N.: Consortium Blockchain for security and privacy-preserving in E-government systems. arXiv.Org (2020)
2. Horwitz, J., Lynn, B.: Toward Hierarchical Identity-Based Encryption, pp. 466–481. Springer, Berlin/Heidelberg (2002)
3. Sandhu, R.S., Coyne, E.J., Feinstein, H.L., Youman, C.E.: Role-based access control models. Inst. Elect. Electron. Eng. (IEEE). **29**(2), 38–47 (1996)
4. Zhang, Y., Deng, S., Zhang, Y., Kong, J.: Research on government information sharing model using Blockchain technology. In: 2019 10th International Conference on Information Technology in Medicine and Education (ITME) (2019)
5. Liu, Y., Zeng, J.: Government data sharing based on Blockchain. In: 2021 The 3rd International Conference on Blockchain Technology (2021)
6. Elisa, N., Yang, L., Chao, F., Cao, Y.: A framework of blockchain-based secure and privacy-preserving E-government system. Wirel. Netw. (2018)
7. Piao, C., Hao, Y., Yan, J., Jiang, X.: Privacy preserving in blockchain-based government data sharing: a Service-On-Chain (SOC) approach. Inf. Process. Manag. **58**(5), 102651 (2021)
8. Wan, Z., Liu, W., Cui, H.: HIBEChain: a hierarchical identity-based Blockchain system for large-scale IoT. IEEE Trans. Depend. Secure Comput. **20**(2), 1286–1301 (2023)
9. Cruz, J.P., Kaji, Y., Yanai, N.: RBAC-SC: role-based access control using smart contract. IEEE Access. **6**, 12240–12251 (2018)
10. Mitrevska, S., Vrangalovska, E., Baloska, S., Mechkaroska, D., Domazet, E.: Blockchain as a Service, an overview on AWS and its BaaS. In: 2022 30th Telecommunications Forum (TELFOR) (2022)
11. Nguyen, L.-T., Nguyen, L. D., Hoang, T., Bandara, D., Wang, Q., Lu, Q., Xu, X., Zhu, L., Popovski, P., & Chen, S.: Blockchain-empowered trustworthy data sharing: fundamentals, applications, and challenges. arXiv.Org (2023)
12. Chen, T., Yu, Y., Duan, Z., Gao, J., & Lan, K.: BlockChain/ABE-based fusion solution for E-government data sharing and privacy protection. Proceedings of the 2020 4th International Conference on Electronic Information Technology and Computer Engineering (2020)
13. Behnke, K., Janssen, M.F.W.H.A.: Boundary conditions for traceability in food supply chains using blockchain technology. Int. J. Inf. Manag. **52**, 101969 (2020)

Design of an SSL Proxy Cryptographic Application Framework

Lishun Ni

1 Introduction

Securing HTTP services with cryptography for authentication and data safety is crucial but poses challenges. Common practices—SSL VPNs and shifting to HTTPS—have drawbacks: VPNs encounter cloud integration difficulties, struggle with scalability, and are hardware-limited, while app encryption requires significant cryptographic skills. This paper introduces an SSL Proxy Cryptographic Framework, a VPN alternative eliminating app changes, enabling effortless cloud onboarding, load distribution, and fitting diverse cloud setups. It surpasses hardware constraints, ensuring scalable, high-performance SSL services for secure cloud data access and provides a practical and effective solution for secure and efficient use of SSL in cloud platform environments.

2 SSL Proxy Cryptographic Application Framework

The SSL proxy cryptographic application functional framework, as shown in Fig. 1, positions the SSL proxy mechanism in front of the WEB application server, serving as the first point of contact for user terminals [1]. These terminals utilize digital certificates and browser cryptography modules to initially connect to the SSL proxy, which then establishes a bidirectional SSL link with the terminal to SSL proxy identity authentication and ensure encrypted transmission. The WEB application server continues to maintain its HTTP WEB service without requiring

L. Ni (✉)
The 3rd Research Institute of Ministry of Public Security, Shanghai, China
e-mail: nilishun@gaskmp.cn

X. Jia et al. (eds.), *2nd International Conference on Big Data, IoT, and Cloud Computing*, EAI/Springer Innovations in Communication and Computing,
https://doi.org/10.1007/978-3-031-93570-1_8

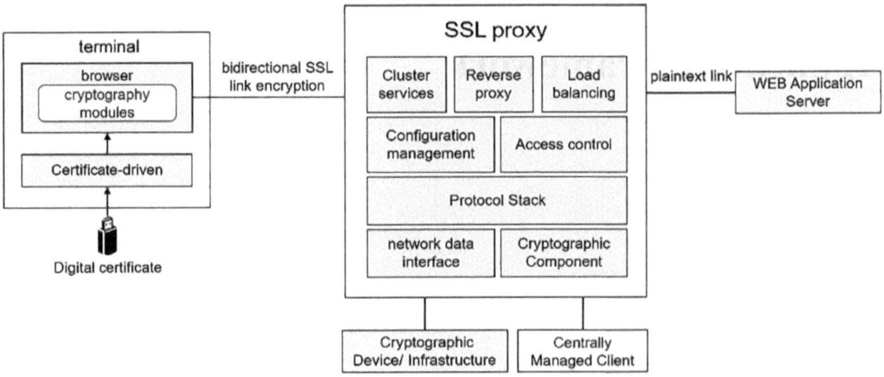

Fig. 1 SSL proxy cryptographic application framework

any modifications. The SSL proxy service leverages cryptographic devices or infrastructure to manage the private key signing for WEB application certificates.

The SSL proxy encompasses several key functionalities, including network data interfaces, cryptographic components, SSL protocol handling, access control, configuration management, cluster services, reverse proxy capabilities, and load balancing.

Network Data Interface: Based on high-speed network cards, this component facilitates rapid transmission and reception of network packets, providing a high-performance network processing foundation for upper-layer functions like protocol stack, access control, and reverse proxy operations.

Cryptography Component: Utilizing cryptographic devices or infrastructure, it provides a set of high-speed encryption application interfaces optimized for SSL protocol services, facilitating signature verification, encryption, and decryption processes. It supports algorithms including RSA, ECDSA, among others [2]. It also incorporates functionality to update SSL session keys based on time intervals or packet traffic volume.

Protocol Stack: Implements parsing and encapsulation services for Ethernet, TCP/IP [3], HTTP, HTTPS, and other protocols, forming the core of the SSL proxy's business processing capabilities.

Access Control: Empowered by the network data interface and protocol stack, it enforces fine-grained policy-based filtering and interception of network packets at the link, network, and application layers, with control granularities down to IP addresses, ports, and protocols.

Configuration Management: Facilitates configuration of operational parameters and policies for the SSL proxy, including cipher suites, certificates, and IP addresses.

Cluster Services: Enables configuration and heartbeat management across multiple SSL proxies, ensuring seamless collaboration and failover within the cluster.

Reverse Proxy: Provides Layer 4 and Layer 7 reverse proxy protection for WEB application services, masking the true IP addresses of application servers from external entities to mitigate security risks, and managing request and response messaging on behalf of the application servers.

Load Balancing: Offers a variety of load distribution methods (e.g., round-robin, IP hashing, weighted distribution) to optimize the distribution of incoming traffic across the WEB application server cluster.

3 High-Performance SSL Proxy Mechanisms

With the advancement of information technology, the processing capabilities of SSL VPNs are pivotal in supporting Web application services. Under traditional models, SSL VPNs are inherently limited by their own hardware performance. In contrast, the SSL proxy service innovatively bifurcates the computational tasks: general functions, such as reverse proxying and protocol encapsulation, which are common and versatile, can leverage universal computing resources for scalability and expansion. Conversely, operations that involve the handling of private keys, specifically actions like signing, are entrusted to dedicated cryptographic devices or cryptographic infrastructure. This strategic separation frees the system from the constraints of specialized hardware's performance limitations, thereby overcoming a crucial bottleneck and enabling a more agile and adaptable security solution that aligns with the evolving demands of information systems.

3.1 Utilizing Zero-Copy Mechanisms for Network Data Retrieval

In the traditional model, once network data reaches the network interface card (NIC), it undergoes a first copy process, moving the data into the operating system's kernel protocol stack. From there, the Virtual File System (VFS) encapsulates this data into a socket, which is then made available to the HTTPS Protocol Service running in user space [4]. Consequently, the data must be copied from kernel space to user space for the gateway program to utilize. This process involves one interrupt per data packet and incurs two separate copying actions, leading to substantial consumption of system resources.

DPDK is a collection of libraries and drivers designed for expedited packet processing and is compatible with multiple platforms including X86 and ARM architectures. This paper leverages the DPDK to implement network data reading and writing for the SSL proxy. DPDK's utilization aims to bypass the conventional Linux kernel path for network I/O, thereby enabling zero-copy operations and significantly reducing the resource overhead mentioned earlier.

Figure 2 shows that the network data interface of the SSL proxy employs the DPDK mechanism, fetching network data directly from the NIC and transferring it

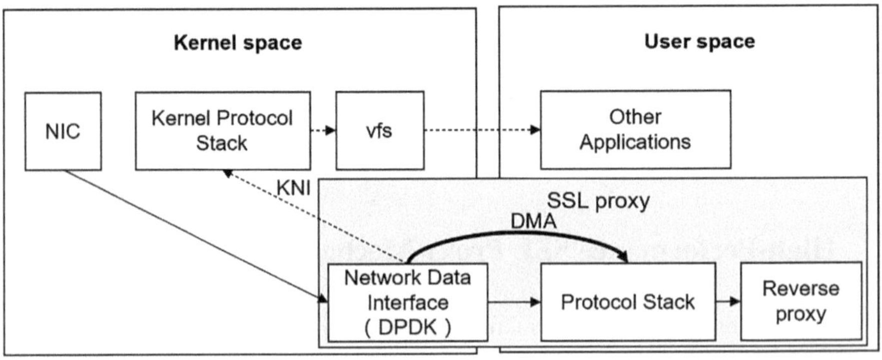

Fig. 2 The DPDK process of reading network data

via Direct Memory Access (DMA) to the user-space protocol stack for parsing and subsequent SSL proxy processing tasks. This methodology eliminates the need for data copying throughout the process, and the DMA operation does not require CPU intervention.

3.2 Employment of Huge Pages Mechanism

Memory blocks in the operating system's kernel protocol stack are typically 4 KB in size, which can lead to excessive page swapping during the parsing of large volumes of network data, negatively impacting performance. The SSL proxy utilizes a huge page's mechanism, with each page measuring 2 MB, supporting more than 512 such pages, translating to direct utilization of over 1 GB of memory. This significantly reduces page swapping frequency and enhances data processing efficiency.

3.3 Support for Multi-queue Network Interface Cards

Modern server-grade NICs often incorporate multi-queue features to balance interrupt loads from each queue across multiple CPU cores, averting single-core saturation while utilizing idle resources through stream distribution to separate CPUs. Retaining streams on dedicated CPUs aligns with TCP's sequential behavior amidst CPU parallelism. This paper's SSL proxy software is designed to support multi-queue NIC technology, fully harnessing the performance potential of the NIC hardware.

3.4 Polling Mechanism

Traditional network cards trigger CPU interrupts for packet receipt or transmission completion, necessitating register saves, ISR execution, and restorations—costly for high-performance networking. Our SSL proxy employs a poll-based approach, bypassing interrupt overhead.

3.5 CPU Affinity

Traditionally, multiple threads share a multicore processor's cores, incurring overhead from frequent context switches that save and restore processor states. CPU affinity optimizes this by assigning threads to specific cores, eliminating switch overhead and cache inefficiencies from core migrations. The SSL proxy employs DPDK with Linux pthreads to bind threads directly to individual CPUs, enabling focused use of isolated resources for enhanced data processing performance.

4 Security Design

In practical applications, the SSL proxy, deployed as an SSL gateway in front of actual web application services, confronts various network protocol attacks, such as unauthorized network probing, DDoS attacks, and the challenge of integrating with existing security devices like firewalls and WAFs, necessitating a holistic security strategy.

To address the issue of unauthorized network probing and scanning, the SSL proxy implements a predefined allow list in formats such as JSON [5], YML, to regulate source-end access. During protocol stack processing, link layer protocol parsing compares the source MAC address against this allow list, discarding packets whose MAC addresses do not match. Similarly, at the network layer, source IP addresses are verified against the allow list, with non-matching packets also discarded or routed to honeypot [6]. These measures effectively render the node invisible to attackers, offering no response.

To combat DDoS attacks, like TCP and UDP Floods, the SSL proxy incorporates traffic cleaning in its protocol handling, dynamically monitors attack rates, and discards packets exceeding set thresholds.

Regarding the deployment strategy with firewalls and WAFs, it is advised that firewalls be positioned externally to the SSL proxy to enforce network access control for incoming traffic. WAFs, on the other hand, can be configured in a bypass mode between the SSL proxy and the web application. Given the HTTP protocol layer between the SSL proxy and the web application, the WAF can effectively intercept and filter HTTP application data for enhanced security.

5 Application Scenarios

The SSL proxy can support a variety of deployment modes, including inline, bypass, and tunneling configurations. Furthermore, it is adaptable to deployment across physical servers, within virtualization containers [7], and in cloud environments [8]. This versatility accommodates diverse infrastructural setups and operational requirements, ensuring compatibility with a broad range of application scenarios.

In cloud platforms, the SSL proxy can serve as a PaaS layer service, providing support for SaaS applications. Its deployment within the cloud takes the form of a Virtual Private Cloud (VPC) structure, as depicted in the referenced diagram. Within the IaaS layer of the cloud platform, virtual network resources are established based on server network interfaces, including a virtual switch and virtual networks (VNETs). These VNETs are mapped to the eth network interfaces of various virtual machines (VMs) in the cloud platform.

As shown in Fig. 3, Vnet1 facilitates communication within VPC1 for VM-node1, VM-node2, and additional VMs. Specifically, VM-node1 hosts the SSL proxy, while VM-node2 and VM-node3 run HTTP applications. The SSL proxy deployed on VM-node1 is instrumental in furnishing SSL protocol services to the other HTTP applications residing within the VPC. This setup enables secure communication for SaaS applications by offloading SSL/TLS processing from the application servers, thereby enhancing security, scalability, and overall performance in the cloud environment.

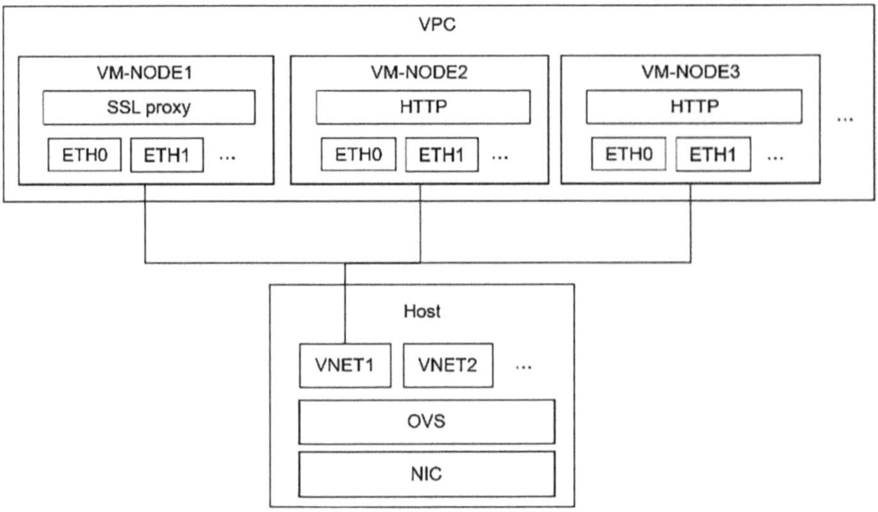

Fig. 3 VM deployment modes of SSL proxy

6 Deployment Verification

To validate the mechanism, we configured an environment (Fig. 4) and ran simulations. The outcomes included successful mutual SSL authentication between the laptop client and proxy, enabling SSL protocol execution and round-robin proxy access to dual HTTP services. A Dockerfile was used to create an SSL proxy container image, facilitating scalable deployment of multiple SSL proxy Docker instances, thus ensuring system flexibility and scalability. Performance tests showed a 26.3% enhancement over standard VPNs.

7 Conclusion

This paper introduces an SSL Proxy Cryptographic Application Framework, designed to supplant traditional SSL VPN appliances, all while circumventing the need for application modifications. The framework is characterized by its support for load balancing, cluster management, on-premises to cloud deployment scenarios, thereby liberating SSL services from the performance constraints inherent in hardware-based VPN devices and delivering high-performance SSL protocol services.

Future work may include researching the implementation mechanisms of integrating SSL proxy within service mesh and sidecar architectures, leveraging these architectures to facilitate more convenient provision of SSL capabilities for cloud platforms.

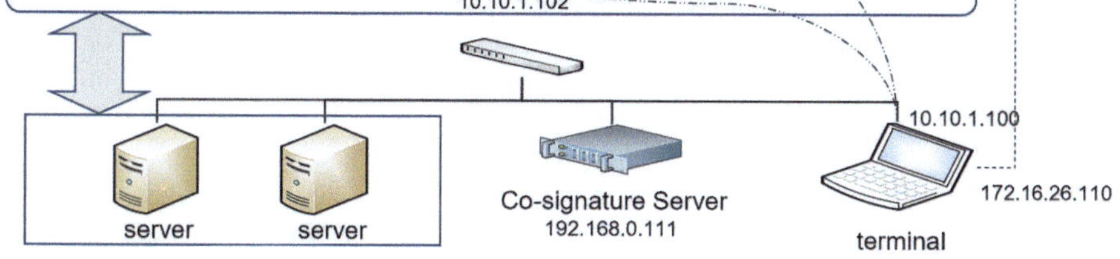

Fig. 4 Validation test deployment diagram

References

1. Jawi, S.M., Ali, F.H.M.: Non-intrusive SSL/TLS proxy implementation and issues [C]. In: Research & Development. IEEE (2016). https://doi.org/10.1109/SCORED.2015.7449426
2. Waziri V, Adebayo O, Danladi H, et al. Network Security in Cloud Computing with Elliptic Curve Cryptography. 2013
3. Kulkarni, H., Pore, R., Agrawal, S., et al.: A survey on TCP/IP API stacks based on DPDK [J]. IJARIIE. **2** (2017)
4. Yuan, D., Wei, H., Zhu, Y.: Design and implementation of user-space protocol stack based on DPDK. J. Inform. Eng. Univ. **24**(1), 93–97 (2023)
5. Jawi, S.M., Ali, F.H.M., Zulkipli, N.H.N.: Nonintrusive SSL/TLS Proxy with JSON-Based Policy [J]. Springer, Berlin/Heidelberg (2015). https://doi.org/10.1007/978-3-662-46578-3_51
6. Zhang, Y.: Design and Implementation of User-Space Network Driver for Routers Based on DPDK. Doctoral dissertation. Shandong University (2019)
7. Song, W., Shen, L., She, W.K.: Research and implementation of high-performance networking module for virtualization systems based on DPDK. Sci. Technol. Innov. Herald. **17**(20), 7 (2020)
8. Patidar P, Bhardwaj A. Network Security through SSL in Cloud Computing Environment [J]. ijcsit.com, 2012

Design of an Efficient Encryption Storage Framework

Lishun Ni

1 Introduction

Amidst rising data volumes and sharing, securing big data storage is imperative. Conventional methods, like investing in costly crypto hardware or isolating data, pose issues: the former is expensive and scalability-limited, while the latter stifles data utility. This paper introduces a high-performance SPDK-based data encryption solution. It efficiently encrypts vast data quantities, enhancing encryption speeds to optimize data processing, achieving a harmony between robust security and operational efficiency, providing a practical and efficient solution for securing the encrypted storage of a large volume of sensitive data.

2 A Cryptography Storage Framework Based on SPDK

The design objective of SPDK revolves around fully exploiting the high-performance potential of contemporary storage hardware, notably NVMe SSDs, by bypassing conventional kernel driver models. By executing storage device driver code in user space, it mitigates overhead associated with system calls and kernel context switches. Notably, while SPDK does not inherently provide encryption or compression functions, it furnishes ample flexibility and interfaces enabling developers to integrate encryption libraries or compression algorithms, thereby achieving efficient and secure data storage. Grounded in these attributes, this paper outlines the design of an SPDK-based encryption storage framework.

L. Ni (✉)
The 3rd Research Institute of Ministry of Public Security, Shanghai, China
e-mail: nilishun@gaskmp.cn

X. Jia et al. (eds.), *2nd International Conference on Big Data, IoT, and Cloud Computing*, EAI/Springer Innovations in Communication and Computing,
https://doi.org/10.1007/978-3-031-93570-1_9

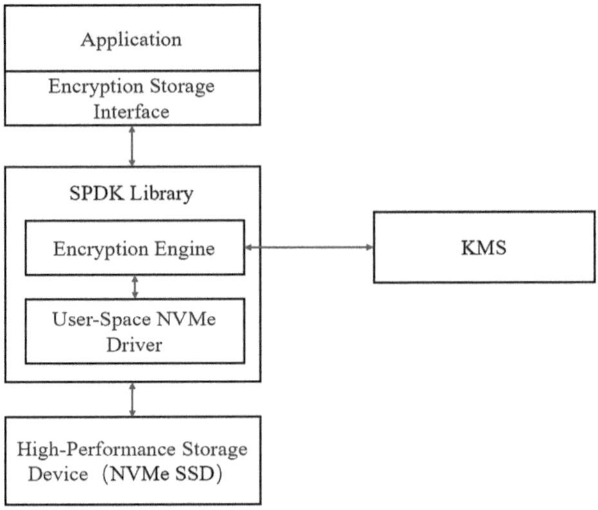

Fig. 1 Encryption storage framework based on SPDK

Figure 1 shows that an application interacting with a storage framework through encrypted storage interfaces. These APIs abstract away complex encryption processes, making them transparent to the application [1]. Key components and their interactions are as follows.

Application Layer: Applications submit read/write requests via encryption-aware APIs that encapsulate encryption logic.

SPDK Layer: User-Space NVMe Driver bypassing the kernel, this driver directly controls NVMe SSDs for high-speed data transfers, leveraging DPDK for efficient user-space I/O handling. Encryption Engine implements various cryptographic algorithms (e.g., SM4, SM3, AES, 3DES) to encrypt data before writing to the SSD and decrypt upon reading.

NVMe SSD: Acts as the high-speed storage medium, benefiting from SPDK's optimized I/O handling for rapid data access [2].

Key Management System: As a standalone, separated component [3], manage the lifecycle of encryption keys, including generation, secure storage, and regular rotation, to ensure the security of the keys is maintained.

The entire architecture is designed to leverage SPDK's performance benefits without compromising on data security during storage and transmission. By integrating encryption directly into the I/O path, it safeguards data against unauthorized access, making it suitable for applications demanding both high performance and stringent security, such as cloud storage solutions and mission-critical business systems.

3 Mechanism Design Implementation

Conventional storage architectures exhibit several characteristic steps and factors that collectively contribute to suboptimal performance, primarily encompassing the following aspects: frequent context switching, interrupt handling, redundant data copying, reliance on small page memory, static configuration, and a lack of flexibility. In contrast, this paper proposes a high-performance framework for encrypted storage based on SPDK, primarily focusing on minimizing system overheads, optimizing data pathways, and ensuring efficient execution of encryption operations. The detailed design is outlined as follows.

User-Space Drivers and Zero-Copy Techniques: By deploying user-space drivers that operate hardware directly from user space, SPDK circumvents kernel context switches and costly system calls, dramatically reducing latency. Coupled with zero-copy methodologies enabled by memory mapping and direct DMA/UPM-DMA [4], data transfer between application memory and storage media is streamlined, eliminating the need for CPU-intensive data replication.

Asynchronous Parallel Processing: The fully asynchronous I/O model of SPDK facilitates concurrent handling of numerous non-blocking I/O requests, capitalizing on multi-core CPUs to minimize I/O idle times. Through multi-threading and an efficient polling mechanism, it supplants conventional interrupt-driven approaches, further decreasing latency.

Memory Management and Caching Strategies: Adoption of large page memory mitigates TLB misses, thereby boosting memory access speed, particularly crucial for intensive read-write operations. Furthermore, implementing intelligent caching strategies, including LRU replacement policies and differentiated read/write caches, reduces direct accesses to storage media, accelerating response times.

In summary, the high-performance encrypted storage framework based on SPDK achieves a dual guarantee of storage performance [5] and data security through profound hardware driver optimizations, efficient software designs, and robust key management mechanisms.

The proposed framework is also tailored to enhance I/O performance for local encrypted storage within virtual machines (VMs), as shown in Fig. 2. Specifically, it leverages SPDK vhost as a backend for VM storage, running on the host machine. This mechanism employs a polling strategy to fetch IO requests from a shared Ring Buffer originating from the VM, processes these requests, and subsequently notifies the VM of the completed operations via the same Ring Buffer mechanism. This end-to-end process is lock-free and devoid of interrupt-induced overheads, thereby achieving remarkable efficiency.

Frontend Driver within the VM: Depending on the employed protocol, this could be virtio blk, virtio scsi, or even the native NVMe driver [6]. Its primary responsibilities include receiving IO requests from the VM, storing them in the Ring Buffer for backend processing, and managing completion notifications from the backend.

Shared Ring Buffer: Acts as a conduit for data, commands, and execution results exchange between the frontend VM driver and the backend hosted on the hypervisor. It is a shared memory region that enables seamless communication without incurring the overhead of traditional inter-process communication mechanisms.

SPDK vhost Backend: Constitutes the core logic for actual IO request handling, functioning as the VM storage backend. It harnesses user-space storage drivers

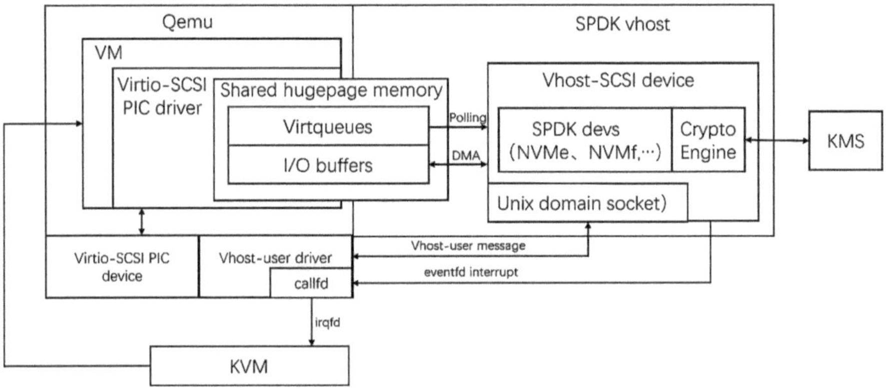

Fig. 2 SPDK encryption storage support for VM

for highly efficient IO operations and integrates encryption engines to facilitate transparent data encryption and decryption [7].

Traditionally, IO requests within VMs traverse extensive call stacks, compounded by duplicative processing steps both within the VM and on the host, leading to increased latency and reduced throughput that negatively affect overall application performance. The SPDK+vhost solution mitigates these inefficiencies by drastically abbreviating the IO processing path inside VMs, thereby attaining low-latency, high-IOPS performance.

4 Advantage Analysis

In the context of the contemporary era of big data, encrypted storage, as a pivotal technology safeguarding data security, has a direct impact on the overall effectiveness of data processing systems. Particularly, the encryption storage mechanism based on SPDK showcases multidimensional advantages. The following is a comprehensive, scholarly overview of the strengths of this system:

Performance optimization and acceleration SPDK achieves revolutionary enhancements in storage performance by relocating critical storage stack components to user space, employing zero-copy techniques, and adopting an asynchronous I/O model. This innovative design avoids the overhead of kernel context switches, and using a polling-driven mechanism, ensures continuous and efficient data transmission.

Enhanced security and compliance under the SPDK framework, encrypted storage not only secures data at rest but also bolsters the safety of data in I/O through advanced key management systems. This transparent encryption mechanism ensures data integrity and confidentiality, thereby providing robust protection for sensitive information. Furthermore, dynamic key management and distribution strategies fortify the security perimeter around key lifecycle management, mitigating the risk of key exposure.

System flexibility and scalability SPDK s modular design and standardized interfaces foster high degrees of flexibility and scalability in systems, enabling developers to easily integrate and configure storage solutions according to specific needs. The architecture is compatible with various storage devices and supports seamless integration into existing IT infrastructures, providing a smooth transition path for system upgrades. Particularly in resource-intensive applications, its horizontal scalability becomes notably evident, where a straightforward addition of hardware resources facilitates linear growth in system performance, thereby accommodating the expansion of future storage demands.

Efficient resource utilization and energy efficiency improvement through the effective employment of large page memory, SPDK optimizes memory access patterns, reducing Translation Lookaside Buffer (TLB) misses and thereby enhancing data processing efficiency. Coupled with intelligent caching strategies, this alleviates pressure on direct storage medium access, decreasing the overall energy consumption of the system. Vital for constructing environmentally friendly and sustainable data center ecosystems, this also leads to a substantial reduction in long-term operational costs.

5 Verification Testing

To comprehensively assess the performance and security of the encryption storage framework based on SPDK, a series of rigorous test schemes have been designed with the aim to validate its effectiveness and advantages in practical application scenarios. This section outlines the key elements of the test design, encompassing performance benchmarking, verification of encryption efficiency, security assessment, and examination of system scalability.

Performance Benchmarking: Verify the performance differential of the SPDK encryption storage framework when encryption is enabled, compared to both an unencrypted state and conventional storage solutions. Employ the Flexible I/O Tester (FIO) tool to simulate diverse I/O workloads, including random read/write, sequential read/write, and mixed read/write operations; metrics such as throughput (IOPS), latency, and bandwidth are measured [8]. The SPDK framework should maintain high IOPS and low latency even under encryption, markedly reduces CPU utilization for encryption tasks, with the framework exhibiting approximately a 23% performance improvement over purely software-based encryption solutions.

Security Assessment: Validate the efficacy of data encryption and the security of the key management system. Attempt to breach encrypted data using methods such as known-plaintext attacks and brute-force techniques; scrutinize the key lifecycle processes, including generation, storage, distribution, and revocation, for adherence to best security practices. Confirm that the encryption algorithm is resilient to cracking attempts, and the key management is flawless, ensuring data security.

Examination of System Scalability: Evaluate how system performance and resource utilization evolve with increasing storage scale and workload. Gradually increase the number of storage devices and I/O loads, monitoring system performance indicators like throughput and latency alongside resource usage (CPU, memory). The system should scale linearly, maintaining robust performance characteristics and efficient resource utilization, thereby validating its stability and reliability in large-scale deployments.

6 Conclusion

This article introduces a high-performance encryption framework based on SPDK, which supersedes traditional file encryption methods and standalone hardware-encrypted disks. Unlike these conventional approaches, it enables encryption without necessitating application modifications and is adaptable to a broad range of scenarios, including big data applications, virtualization, and cloud storage. This framework transcends the performance and cost constraints of previous solutions, delivering a robust, high-performance encrypted storage service.

References

1. Mingwei, Z., Rui, M., Rongan, J.: A transparent encryption file system model based on filter driver [J]. Comput. Eng. **35**(1), 3 (2009). https://doi.org/10.3969/j.issn.1000-3428.2009.01.051
2. Zou, Y., Awad, A., Lin, M.: DirectNVM: hardware-accelerated NVMe SSDs for high-performance embedded computing. ACM Trans. Embed. Comput. Syst. **21**(1), 1 (2022)
3. Zhizhong, W., Cheng, Z., Yufei, M.: A cloud storage encryption solution based on separated keys [J]. Telecom. Sci. **29**(1), 51–56 (2013). https://doi.org/10.3969/j.issn.1000-0801.2013.01.008
4. Zhu, J., Xiao, L., Wang, L., et al.: UPM-DMA: an efficient user space DMA-pinned memory management strategy for NVMe SSDs [C]. In: International Conference on Algorithms and Architectures for Parallel Processing. Springer, Cham (2022). https://doi.org/10.1007/978-3-030-95384-3_17
5. Yang, Z., Liu, C., Zhou, Y., Liu, X., Cao, G.: SPDK Vhost-NVMe: accelerating I/Os in virtual machines on NVMe SSDs via user space Vhost target. In: 2018 IEEE 8th International Symposium on Cloud and Service Computing (SC2). IEEE (2018)
6. Kim, H.J., Kim, J.S.: A user-space storage I/O framework for NVMe SSDs in mobile smart devices. IEEE Trans. Consum. Electron. **63**(1), 28–35 (2017). https://doi.org/10.1109/TCE.2017.014709
7. Yang, Z., Harris, J.R., Walker, B., et al.: SPDK: a development kit to build high performance storage applications [J]. IEEE. (2017). https://doi.org/10.1109/CloudCom.2017.14
8. Xu, Q., Siyamwala, H., Ghosh, M., et al.: Performance analysis of NVMe SSDs and their implication on real world databases [J]. ACM, 1–11 (2015). https://doi.org/10.1145/2757667.2757684

Secure Cryptography Framework for Cloud Applications

Lishun Ni

1 Introduction

As information systems migrate to the cloud, they require robust cryptography to protect sensitive data. Existing cryptographic systems, however, struggle to fully integrate with cloud dynamics, such as scalable VMs, secure access within software-defined networks, effective deployment of diverse cryptographic devices in cloud environments, and ensuring safe key management in applications. And homomorphic encryption is yet insufficient to meet application requirements [1]. This paper introduces a novel, loosely coupled framework designed specifically for cloud platforms. It tackles these challenges head-on, improving the security and flexibility of cryptographic processes in the ever-evolving cloud ecosystem. This framework supports seamless cryptography integration and adaptive management, enhancing cloud security practices.

2 Design of Cryptographic Application and Management Framework

The design of a loosely coupled cryptographic application and management framework with cloud platforms is shown in Fig. 1. This framework primarily consists of cryptographic infrastructure that can be deployed independently of cloud platforms

L. Ni (✉)
The 3rd Research Institute of Ministry of Public Security, Shanghai, China
e-mail: nilishun@gaskmp.cn

X. Jia et al. (eds.), *2nd International Conference on Big Data, IoT, and Cloud Computing*, EAI/Springer Innovations in Communication and Computing,
https://doi.org/10.1007/978-3-031-93570-1_10

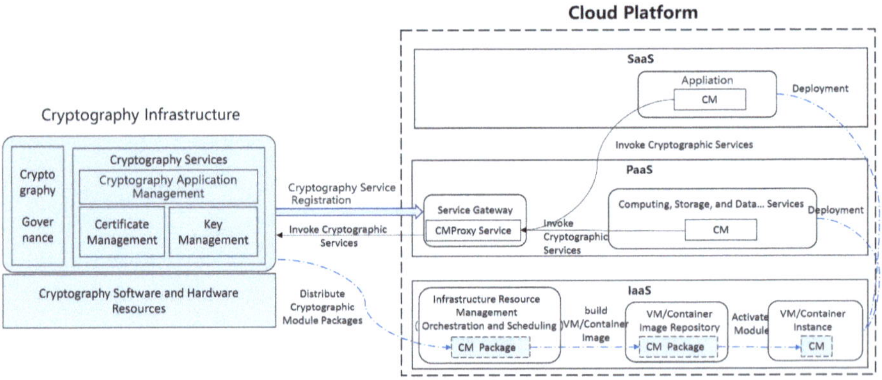

Fig. 1 Cryptographic application and management framework

[2] and cryptographic modules implemented as software middleware within the cloud platform.

The cryptographic infrastructure is deployed based on independent resources for computation, storage, networking, security, and cryptographic devices. It provides capabilities such as certificate management, key management, and cryptographic services. Certificate management encompasses the full lifecycle of certificates, including issuance, revocation, reinstatement, updating, while also implementing registration services, directory services, and online query services for certificate status [3]. Key management builds upon mechanisms like key dispersion and derivation to construct a symmetric key system, managing the entire lifecycle from key generation, storage, distribution, backup, archiving, to destruction [4]. Cryptographic services, leveraging issued certificates and distributed keys, employ secure cryptographic algorithms to deliver various cryptographic services to applications. The cryptographic application management, as a submodule of cryptographic services, systematizes management of cryptographic modules within applications, configures cryptographic policies, and facilitates cryptographic supervision. The components for certificate management, key management, and cryptographic services are encapsulated as independent units, outputted as services, forming a layered, decoupled, flexible, scalable, and reliable cryptographic infrastructure.

The cloud platform is segmented into Infrastructure-as-a-Service (IaaS), Platform-as-a-Service (PaaS), and Software-as-a-Service (SaaS) layers. The diagram denotes the resources and services at each layer relevant to this framework. At the IaaS layer, basic resource management handles the orchestration and scheduling of VMs/containers, with the VM/container repository storing their images and instances representing instantiated VMs/containers. The PaaS layer's service gateway manages service registration and access, while fundamental services like computing, storage, and data are provided to applications as platform services. At the SaaS layer, applications leverage VM/container instances and platform services to implement diverse functionalities.

The cryptographic infrastructure's application management enables administrators to create and distribute installation packages for registered cloud applications' cryptographic modules. Through agreed protocols such as SFTP or network file systems, the cloud platform's basic resource management acquires these cryptographic module packages, integrates them into VM/container image files, and archives them in an image repository. When deploying VM/container instances for cloud applications, the platform retrieves images from this repository, installs and activates the cryptographic modules. Consequently, when cloud applications utilize VMs/containers, these come pre-equipped with activated cryptographic modules. Applications then invoke local SDKs provided by the cryptographic modules to access cryptographic services from the infrastructure, thus accomplishing cryptographic.

The entire framework operates without the need to host hardware cryptographic devices within the cloud infrastructure; it merely requires that the cryptographic infrastructure be network-accessible from the cloud platform.

2.1 Security Control of Cryptographic Modules

Cryptographic module management handles registration to de-registration, using app and cloud resource details for package creation and activation verification. Activation includes reporting runtime info, verifying it against registration data, and securing activation via OTPs/pre-configured keys, adaptable to the app's security level with automated or hybrid approval process.

Modules activate, generate keys via co-signing with the cryptographic infrastructure, and obtain a certificate for subsequent authentication. Heartbeat info and runtime checks ensure module integrity, triggering alerts and actions for anomalies. Runtime defenses like code signing, obfuscation, and memory randomization, plus cloud security features, protect modules from tampering and unauthorized access.

Collaborative signing [5] enables TLCP and mutual authentication between modules and infrastructure, reinforcing secure communication and preserving the cryptographic operation's integrity and confidentiality. This integrated approach safeguards the entire lifecycle of cryptographic modules within the cloud environment.

2.2 Secure Cryptographic Management

Cloud application cryptography revolves around managed cryptographic modules. Applications register with the crypto infrastructure to gain access to setup packages, certificates, keys, and services. Sensitive keys are securely stored within infrastructure devices, accessed only via authenticated module certificates post-deployment. This enforces strict access control for secure key usage.

The infrastructure updates keys and certs regularly, maintaining security. Any module issues necessitate restoration before applications can resume cryptographic functions. Applications provide initial info and module requirements during a registration process compatible with the cloud platform's systems, manual or automated, without limiting scalability.

The infrastructure centrally manages a lot of cryptographic modules, streamlining crypto resource use, governance, and ensuring policy adherence. This design enables efficient, secure, and scalable crypto services, meeting the diverse needs of cloud applications while enhancing overall cryptographic security and oversight.

2.3 Cryptographic Service Support

The cryptographic application and management framework ensures secure and easy use of cryptographic tech for tasks like authentication and data encryption [6]. It achieves this via a centralized cryptographic infrastructure supporting various cloud application scenarios.

Applications authenticate via cryptographic modules and a secure channel to access services. Small data encryption involves secure data transfer, using app-specific keys and algorithms. For large data, local random numbers act as one-time pads, with keys encrypted using app certificates, minimizing external data transfers for efficiency.

Digital signatures are generated by the cryptographic module, which performs a hash operation on the signing data. The hash value is transmitted via the secure channel to the cryptographic service, which uses the application's private key certificate and a specified algorithm to complete the signature. Verification involves fetching public keys, offloading compute-intensive tasks to local modules [7].

For unified services like electronic seals, applications access them remotely via module-provided proxies after secure authentication.

The framework integrates these mechanisms, enabling applications to swiftly adopt more cryptographic scenarios through module interfaces, enhancing service efficiency and flexibility in diverse applications.

2.4 Cryptographic Application Supervision

Cryptographic application supervision holds pivotal significance in ensuring the effectiveness of cryptographic operations by providing overarching guidance and management. This framework capitalizes on the advantages of a centrally built cryptographic infrastructure and the deployment of cryptographic modules within cloud-based applications. On the one hand, it gathers management and operational logs from key management, certificate management, cryptographic application management, and various cryptographic services within the cryptographic infrastructure

[8]. On the other hand, cryptographic modules report their operational logs and logs detailing application invocations of the cryptographic modules back to the cryptographic infrastructure. By analyzing these dual sources of data, a comprehensive, precise, and detailed understanding of the actual usage of cryptographic services by cloud applications can be obtained. Furthermore, integrating this analysis with application business logs and cloud platform operational logs allows for a holistic overview of cryptographic application usage in the cloud. This overview enables administrators to gain a concrete understanding of the keys, algorithms, and volumes of cryptographic operations employed by different application services, facilitating the formulation of practical strategic recommendations that enhance the level and effectiveness of cryptographic management and application.

3 Advantage Analysis

This innovative cryptographic framework addresses security, platform coupling, cloud compatibility, and regulatory oversight shortcomings in existing solutions. It fortifies security by employing dedicated hardware for key management, using secure storage, and verified access, thus mitigating risks associated with insecure software practices. Unlike tightly coupled systems, it ensures loose coupling with cloud platforms through a modular, network-accessible design compatible across various cloud ecosystems, supporting seamless integration and operation within cloud-native architectures.

The framework embraces standard URL protocols for cryptographic services, allowing straightforward registration with PaaS layers and direct interaction through service gateways or software-defined networks. It comprehensively supports essential cryptographic functions, packaging them into intuitive, scenario-specific interfaces for easy adoption by developers, thereby enhancing usability without requiring deep cryptographic knowledge.

Regarding supervision, it centralizes cryptographic resources and log management, streamlining regulatory compliance. Cryptographic modules not only provide local service interfaces but also collect and report usage logs uniformly, alleviating the burden on individual applications. This consolidated approach to logging and oversight simplifies data collection and significantly bolsters the efficiency of cryptographic monitoring, ensuring a robust, flexible, and secure environment for cryptographic applications in diverse cloud settings.

Centralized construction of cryptographic infrastructure enables holistic management of cryptographic devices, allowing for rational utilization of resources and enhanced efficiency. Furthermore, by combining cryptographic modules with the cryptographic infrastructure to form a cohesive cryptographic capability, as detailed in Sect. 2.3, the performance of cryptographic services can be significantly improved. This approach holds a performance advantage over traditional hardware-intrusive methods.

4 Deployment Verification

To test the proposed framework, we configured a cloud-native development environment with two servers, a switch, OpenStack, Kubernetes, and Docker for IaaS, GitLab, Jenkins, Harbor, Spring Cloud Gateway, and Nacos were deployed as PaaS services, managing code, builds, Docker images, service gateway, and registration respectively, as shown in Fig. 2. A default Spring Cloud app ran atop Docker as a mock SaaS layer.

A separate cryptographic infrastructure was set up with another server hosting an HSM and virtual machines for cryptographic services, including an SFTP for module distribution. These systems were network-isolated yet interconnected, reflecting real-world segmentation.

In the simulation, Kubernetes scaled down existing containers and added new ones. Service tests verified sustained HTTPS support and smooth encryption-decryption integration post-scaling. Thus, the validation confirmed a non-intrusive, scalable, and loosely coupled cryptographic framework, adhering to cloud-native principles, and achieving its design goals.

5 Conclusion

Current cryptographic systems fall short in adapting to cloud demands, including scalable VM/container management, secure network access, diverse cryptographic device hosting, and secure key handling within cloud applications. This paper

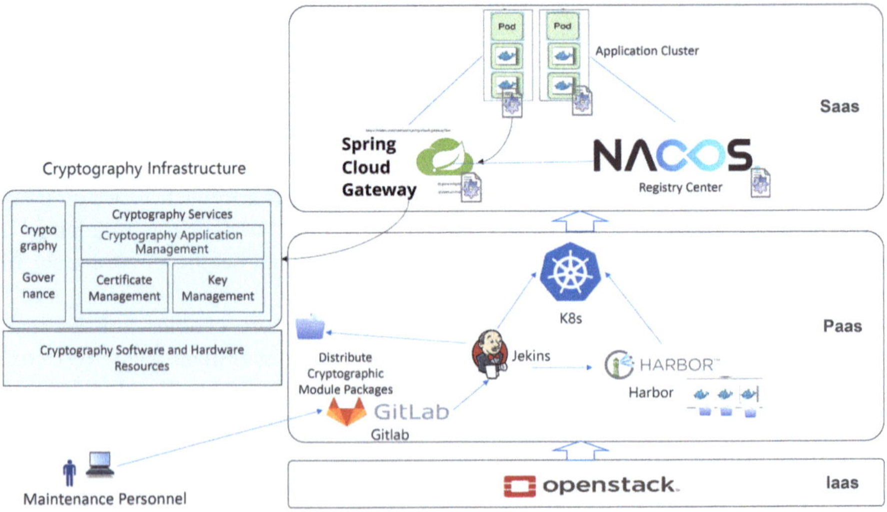

Fig. 2 Deployment of cryptographic application management in cloud-native environment

introduces a loosely coupled framework for cloud platforms, creating an external cryptographic infrastructure integrated into cloud-native applications via hybrid software-hardware, ensuring security, usability, and scalability without platform constraints.

The framework emphasizes full lifecycle management of cryptographic modules, robust security controls, seamless cloud integration, and effective supervision. It excels in providing high-security cryptography within clouds, user-friendly application support, universal supervision, cross-platform compatibility, and flexible app scaling.

Validation involved setting up a test environment with OpenStack, Kubernetes, Docker, and other tools, simulating cryptographic app deployment, scaling, and contraction processes to demonstrate the framework's effectiveness in real-world cloud scenarios.

Future work may include researching the implementation mechanisms of integrating cryptographic modules within service mesh and sidecar architectures, leveraging these architectures to facilitate more convenient provision of cryptographic capabilities for cloud platforms.

References

1. Patra, G.K., Chakraborty, N.: Securing cloud computing environment with the help of fully homomorphic encryption. J. Comput. Technol. Appli. **4**(3) (2013) ISSN: 2229-6964
2. Patra, G.K., Chakraborty, N.: Securing cloud infrastructure for high performance scientific computations using cryptographic techniques. Int. J. Adv. Comput. Res. **4**(14) (2014)
3. Ijaz, I., Aslam, A., Bukhari, B., et al.: Securing cloud infrastructure through PKI [J]. IEEE. (2014). https://doi.org/10.1109/ICCCNT.2014.6963055
4. Livingston, D., Kirubakaran, E., David, E.P.: Implementing data privacy of cloud data on a remote server using symmetric cryptographic algorithms [J]. Int. J. Adv. Data Inform. Syst. **2**(1), 62–72 (2021). https://doi.org/10.25008/ijadis.v2i1.1217
5. Shengwei, X., Ye, D., Yu, T.: Design and application of a collaborative signing scheme based on SM2 for both parties in the cloud [J]. J. Beijing Elect. Sci. Technol. Instit. **31**(1), 1–8 (2023)
6. Manager, E.P.: Lee, A. Cryptographic Key Management (CKM) Design Principles for the Advanced Metering Infrastructure (AMI).
7. Shinde, M.S.: Authentication & Isolation of Cryptographic Services for Cloud Computing [J] (2018). https://doi.org/10.1109/ICSTM.2015.7225417
8. Darne, A., Longadge, R.: A result analysis on secured encryption in cloud computing using symmetric cryptography. Int. J. Comput. Sci. Mobile Comput. **3**(7) (2014)

Multi-Party Computation Scheme Based on the SM2 Algorithm

Hao Liang

1 Introduction

In the context of rapid IT development, the application of domestic cryptographic algorithms like SM2 has become crucial. Software cryptographic modules providing SM2 functionalities are essential where dedicated cryptographic hardware is unavailable. However, storing SM2 private keys in software poses significant security risks. Traditional methods, such as threshold algorithms [1, 2], split the private key to disperse risk but increase complexity and exposure during recombination. Two-party cooperative mechanisms [3–8] enhance security without full key exchanges but are limited to two parties and face synchronization issues. To address these challenges, we propose a multi-party computation scheme based on SM2. This scheme supports multiple participants in signing and decryption, reduces private key leakage risks through distributed storage, and includes a password-based key derivation mechanism for added protection. It ensures secure private key management in software cryptographic modules, expanding the application of domestic cryptographic algorithms to meet growing information security demands.

2 Logical Architecture

In the logical architecture of this scheme, as shown in Fig. 1, the system consists of a scheduling module and multiple key modules. The scheduling module is located

H. Liang (✉)
The 3rd Research Institute of Ministry of Public Security (Key Laboratory of Information Network Security, Ministry of Public Security), Shanghai, China
e-mail: lianghao@gaskmp.cn

X. Jia et al. (eds.), *2nd International Conference on Big Data, IoT, and Cloud Computing*, EAI/Springer Innovations in Communication and Computing,
https://doi.org/10.1007/978-3-031-93570-1_11

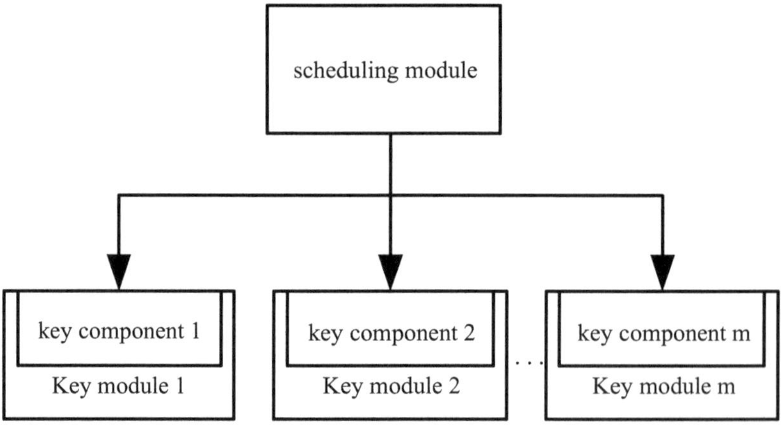

Fig. 1 Logical Architecture

at the center of the system, responsible for coordinating and managing each key module. These key modules can be different submodules within the same system or different subsystems distributed across the network.

Each key module is independently responsible for generating, storing, and managing its respective SM2 algorithm private key components. These key modules are equipped with secure random number generators to produce random SM2 private key components and temporary data k used in encryption operations. Each key module has strict security mechanisms to ensure that private key components cannot be unauthorizedly exported or copied. Private key components are strictly confined for use within the key modules, only performing encryption or decryption operations on externally received data and outputting the results.

The scheduling module does not store any key information. It is responsible for receiving data from external applications and invoking different key modules in a predetermined sequence. During digital signing and decryption operations, the scheduling module transmits the required process data but does not involve any key content to ensure the security of the keys.

This distributed architecture design ensures that each private key component is generated, stored, and processed in its respective independent module. Even if one module is compromised, the private key components stored in other modules remain secure and intact. Such a design significantly enhances the overall security of the system, as an attacker would need to simultaneously breach all key modules to obtain the complete private key. This difficulty increases exponentially with the number of modules, making the system highly secure in practice.

3 Key Generation

The main process of key generation includes:

1. Determine the number of key modules as m. Key module 1 uses a random number generator to generate the first private key component $d_1 \in [1, n-1]$.
2. Key module 1 computes the inverse of d1 in the finite field Fq as d_1^{-1} mod n.
3. Key module 1 computes $P_1 = d_1^{-1}[*]G$.
4. Key module 2 uses a random number generator to generate the second private key component $d_2 \in [1, n-1]$.
5. Key module 2 computes the inverse of d2 in the finite field Fq as d_2^{-1} mod n.
6. Key module 2 computes $P_2 = d_2^{-1}[*]P1[*]G$.
7. Each key module repeats the above steps in sequence until key module $m - 1$.
8. Key module m generates the final private key component dm, computes $P_m = d_m^{-1}[*]P_{m-1}[-]G$.
9. The final common public key FP is published by the last key module, where $FP = P_m$.

During the generation of the common public key, the private key components of each key module are securely kept and are only processed within their respective key modules. They are never combined externally, effectively avoiding the risk of the entire private key being leaked.

The common public key FP corresponding to the private key components of each key module is based on the chain multiplication result of the inverses of the private key components from all key modules. The mathematical expression is $FP = (d_m^{-1} * \ldots * d_2^{-1} * d_1^{-1} - 1)[*]G$. The FP computed in this manner is mathematically equivalent to the public key $P = d[*]G$ corresponding to a single private key d in the SM2 algorithm.

Each key module's output data (denoted as $P_1, P_2 \ldots P_m$) is essentially the result of the point multiplication of the inverses of their private key components with the base point on the elliptic curve. Due to the inherent irreversibility of elliptic curve encryption calculations, even if an attacker obtains these point multiplication results, it is nearly impossible to deduce the plaintext private key components of any module.

After the multi-party key generation process is completed, the resulting common public key FP can be extracted and used for issuing digital certificates through the Public Key Infrastructure (PKI). This digital certificate represents the collective of all key modules involved in forming the common public key. It enables external third parties to verify the legitimacy of signatures generated by this set of key modules or to encrypt data using the public key in the certificate, ensuring that only this set of key modules can decrypt the data. If the private key component of any key module changes or if a private key component is compromised, any signature verification or data encryption using the common public key will no longer be valid. Furthermore, if the digital certificate associated with the common public key is revoked, all private key components of the related key modules become immediately invalid. This mechanism allows the SM2 multi-party key generation scheme to integrate seamlessly with standard cryptographic applications, demonstrating its high practicality and usability.

4 Private Key Signing

As shown in Fig. 2, the main steps of the private key signing process include:

1. The scheduling module inputs the original plaintext M. It computes the SM2 algorithm preprocessing data Z using the signature identifier id, the common public key FP, and the plaintext M. Then, it takes M' = Z||M and computes the message digest e = Hash(M').
2. The scheduling module sequentially calls each key module, generating a random number $k_1 \in [1, n-1]$, computing $Q_1 = k_1[*]G$, generating a random number $k_2 \in [1, n-1]$, computing $Q_2 = k_2 \cdot Q_1$, and so on, until generating the random number $k_m \in [1, n-1]$ and computing $Q_m = k_m [*]Q_{m-1}$. Finally, it computes $r = getX(Q_m) + e \bmod n$.
3. Each key module calls its private key component to compute the sss part of the signature value:

   Compute $s_1 = d_1 * k_1 \bmod n$ and $v_1 = d_1 * r \bmod n$. Compute $s_2 = d_2 * k_2 * s_1 \bmod n$ and $v_2 = d_2 * v_1 \bmod n$. Each module sequentially performs these calculations until module m computes $s_m = d_m * k_m * s_{m-1} \bmod n$ and $v_m = d_m * v_{m-1}$ mod

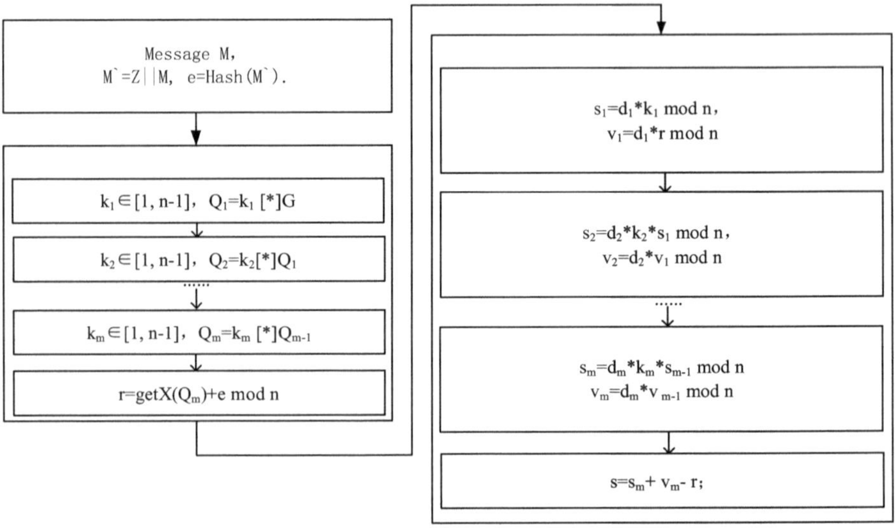

Fig. 2 Multi-party private key signature

n. Compute $s = s_m + v_m - r$. Finally, the scheduling module outputs the signature value (r,s).

This scheme requires all key modules involved in generating the signature to participate, with each contributing part of the information to form the final signature. This design ensures that even if an attacker compromises one key module, they cannot independently generate a valid signature without contributions from all other modules, thus enhancing overall security. During the signing process, each key module independently computes its contribution while maintaining the confidentiality of its private key component. Intermediate variables in the process are results of modular multiplication operations using the private key components, relying on complex mathematical problems in elliptic curve cryptography, such as the discrete logarithm problem. This significantly increases the difficulty of deducing private key components from public information.

To further enhance the security of the signing process, the multi-party private key signing mechanism requires each key module to generate a unique random number k_x and incorporate it into the signing computation. This step not only introduces additional unpredictability but also ensures that even if some information is disclosed, an attacker cannot use this information to reconstruct any party's complete private key. The randomness of each signing process guarantees that the signatures generated with the same private key components are unique and unpredictable each time.

The most important feature of the multi-party private key signing is that the generated signature value is mathematically equivalent to the standard SM2 signature based on a single private key. This means that the common public key produced during the multi-party key generation process can be used to verify the authenticity of the signature.

The calculation of the r part of the SM2 signature value in the standard SM2 algorithm is as follows:

$$r = (x1 + e) \bmod n$$

where x1 is the x-coordinate of the elliptic curve point k, i.e.:

$$r = (x1\,(k\,[*]\,G) + e) \bmod n$$

The Generation Method of the r Part of the SM2 Signature Value in Multi-Party Private Key Signing Operation:

$$r = (x1\,(Q_m) + e) \bmod n$$

$$r = (x1\,(k_m\,[*]\,Q_{m-1}) + e) \bmod n$$

$$r = (x1\,(k_m\,[*]\,k_{m-1}\,[*]\,Q_{m-2}) + e) \bmod n$$

$$r = (x1\,(k_m\,[*]\,k_{m-1}\,[*] \ldots k_1\,[*]\,G) + e) \bmod n$$

Let $k = k_m[*]k_{m-1}[*] \ldots k_1$, substituting into the above formula, we get:

$$r = (x1\,(k\,[*]\,G) + e) \bmod n$$

It can be observed that the r part of the SM2 signature value in the multi-party private key signing operation is consistent with the r part of the signature value in the standard SM2 algorithm.

The calculation of the s part of the SM2 signature value in the standard SM2 algorithm is as follows:

$$s = (1 + d)^{-1} * (k\text{-}r*d) \bmod n$$

In the multi-party private key signing operation, the s part of the SM2 signature value is generated as follows:

$$s = s_m + v_m - r$$
$$s = (d_m*k_m*s_{m-1} + d_m*v_{m-1}\text{-}r) \bmod n$$
$$s = (d_m*k_m*\cdots*d_1*k_1 + d_m*\cdots*d_1*r\text{-}r) \bmod n$$

Let $k = k_m{*}k_{m-1}{*} \ldots k_1$, substituting into the above formula, we get:

$$s = (d_m*\cdots*d_1*k + d_m*\cdots*d_1*r\text{-}r) \bmod n$$
$$s = (d_m*\cdots*d_1*k - r*1 + d_m*\cdots*d_1*r) \bmod n$$

Let $1 = (d_m* \ldots *d_1)*(d_m* \ldots *d_1)^{-1}$, and factor out $(d_m* \ldots *d_1)$, we get:

$$s = (d_m*\cdots*d_1)\left(k\text{-}r*(d_m*\cdots*d_1)^{-1} + r\right)$$
$$s = \left(\left(d_m^{-1}*\cdots*d_1^{-1}\right)^{-1}\right)\left(k\text{-}r*\left(d_m^{-1}*\cdots*d_1^{-1}\right) + r\right)$$
$$s = \left(\left(1 + d_m^{-1}*\cdots*d_1^{-1}\text{-}1\right)^{-1}\right)\left(k\text{-}r*\left(d_m^{-1}*\cdots*d_1^{-1}\text{-}1\right)\right)$$

Substituting $d = d_m^{-1}* \ldots *d_1^{-1}\text{-}1$, we get:

$$s = (1 + d)^{-1} * (k\text{-}r*d) \bmod n$$

This result is consistent with the s part of the signature value generated by the standard SM2 algorithm.

From the above derivation, it can be seen that the multi-party private key signing mechanism essentially transforms the inverse of the single private key d in the standard SM2 algorithm into the cumulative product of the inverses of the private key components held by multiple key modules, minus one. It also transforms a single random number k into the cumulative product of random numbers generated by multiple modules. This method allows the scheduling module to flexibly adjust the order of modules participating in the computation while ensuring that each key module participates in the calculation, improving the system's flexibility and efficiency. When a particular key module is busy, the scheduling module can schedule other idle modules to perform the computation, thus optimizing resource usage and enhancing overall performance.

5 Private Key Decryption

The main steps of the multi-party private key decryption process include:

1. The scheduling module parses the ciphertext data C and extracts C1 from the ciphertext structure.
2. Key module 1 uses its private key component d1 to compute $T_1 = d_1^{-1}[*]C1$.
3. Key module 2 uses its private key component d2 to compute $T_2 = d_2^{-1}[*]T_1$.
4. Each key module sequentially performs the calculation until key module mmm computes $T_m = d_m^{-1}$ $[*]T_{m\text{-}1}$.
5. The scheduling module computes $(x_m, y_m) = T_m[-]C1$.
6. The scheduling module computes $t = KDF(x_m || y_m, klen)$.
7. The scheduling module extracts C2 from the ciphertext and computes $M` = C2 \oplus t$.
8. The scheduling module computes $C3` = Hash(x2||M`||y2)$ and compares it with C3 extracted from the ciphertext. If they are equal, $M'$is output as the decrypted result.

This scheme ensures that data encrypted with the common public key generated through multi-party key generation can only be decrypted when all related private key components are involved. This mechanism significantly enhances system security because it requires the participation of all private key holders in the decryption process. Since the decryption process requires the combined effort of all private key components, even if one private key component is leaked, the data cannot be decrypted independently, providing stricter security guarantees in a multi-party collaborative environment.

The intermediate data T_m transmitted during the multi-party private key decryption process relies on elliptic curve cryptography point multiplication operations. Even if these point multiplication results are intercepted during data transmission, attackers would find it difficult to use this information for an effective attack. One of the core challenges of elliptic curve cryptography is that, even knowing the point multiplication result on the elliptic curve, reversing the multiplier remains computationally infeasible.

The multi-party private key decryption mechanism is operationally equivalent to the standard SM2 algorithm, allowing external entities to encrypt data with a standard SM2 public key without any modifications. Any ciphertext generated with the SM2 public key can be decrypted using the combined efforts of the private key components. This ensures enhanced security and seamless integration with existing encryption infrastructure, requiring no adjustments to the current system architecture or communication protocols. Mathematically, multi-party decryption mirrors single private key decryption, represented as (x, y) = d[*]C1, where d is the private key and C1 is the encrypted data. In multi-party decryption, the private key d is the product of multiple components, expressed as $d = d_{m-1} * d_{m-2} * \ldots * d_1$. When these components jointly perform the decryption, the result is equivalent to using the single private key d, ensuring compatibility and practicality.

6 Conclusion

The proposed SM2 multi-party computation scheme successfully implements a collaborative mechanism for data encryption, signing, and decryption through distributed management of private key components and the generation of a common public key, effectively addressing the challenge of secure private key storage.

In terms of security, the scheme significantly reduces the risk of overall key theft by avoiding centralized storage of private keys. Experimental results show that the system's security increases exponentially with the number of participants, providing a more reliable guarantee for private key management.

Additionally, this scheme maintains operational equivalence with the standard SM2 algorithm, ensuring seamless compatibility with existing secure communication protocols. This characteristic not only simplifies system integration but also offers an opportunity for seamless upgrades in private key management.

References

1. Ming, S., Yuan, M., Jingqiang, L., et al.: SM2 elliptic curve threshold cryptography algorithm. J. Cryptograp. **1**(02), 155–166 (2014). https://doi.org/10.13868/j.cnki.jcr.000015
2. Zeyan, T., Wenjun, L., Xiaofang, H.: Design and implementation of a Mobile terminal signature system based on verifiable SM2 threshold algorithm. Comput. Meas. Control. **27**(03), 225–230 (2019). https://doi.org/10.16526/j.cnki.11-4762/tp.2019.03.046
3. Jinquan, T., Liang, H.: A TLCP secure communication scheme based on SM2 homomorphic signature. China Computer Federation. In: Proceedings of the 38th National Computer Security Academic Exchange Conference. The Third Research Institute of the Ministry of Public Security, p. 4 (2023). https://doi.org/10.26914/c.cnkihy.2023.035896
4. Qi, F., Debiao, H., Min, L., et al.: Lightweight SM2 two-party collaborative signature in Mobile internet environment. J. Comput. Res. Dev. **57**(10), 2136–2146 (2020)
5. Kai, L.: Design and implementation of a two-party SM2 signature scheme. Zhejiang Gongshang University (2019). https://doi.org/10.27462/d.cnki.ghzhc.2019.000427
6. Xipan, L.: Collaborative Signature and Decryption Method and System Based on SM2 Algorithm for Communication Parties. Henan Province, Zhengzhou Xinda Jiean Information Technology Co., Ltd (2018-06-01)
7. Jing, W.: Research on secure two-party computation Technology for Mobile Internet. Wuhan University (2021). https://doi.org/10.27379/d.cnki.gwhdu.2021.000877
8. Wang Jing, W., Libing, L.M., et al.: A secure and efficient two-party collaborative ECDSA signature scheme. J. Commun. **42**(02), 12–25 (2021)

Research on Recommendation Technology of Tobacco Formula Materials Using CBA

Qianxu Yang, Sheng Lei, Wei Zhang, Baokun Zhu, Mingzhan Lv, and Bin Zhang

1 Introduction

Tobacco formulation is a process technology that combines various materials in specific proportions according to the requirements of a particular tobacco product [1]. When the brand formulation lacks complete single materials, it is necessary to select suitable single materials from inventory for replacement. Currently, new formulation design and maintenance rely on expert tasters who combine and replace materials based on experience, then use feedback from scoring to change the formulation until it meets brand requirements. This model is a one-dimensional design approach that lacks consideration for inventory, cost, and efficiency, resulting in high tasting costs and long cycles for formulation updates.

Due to the fast market changes and the pressure of new sales methods, specific brands can now reach target consumers more quickly. Through analysis based on big data of tobacco consumption, Guangxi Zhongyan [2] concluded that the precision of material supply utilization has increased by 24%. Therefore, modern requirements such as digesting tobacco leaf inventory, improving the efficiency of tobacco leaf use, and product digital design [3] have been put forward for tobacco formulation technology.

Early, resolving the compatibility of tobacco leaves and meeting brand sensory requirements by analyzing the chemical components of raw materials was the main method and research direction for maintaining tobacco formulations, such as Wang Jianmin et al. [4] establishing the relationship between chemical components and

Q. Yang · S. Lei · W. Zhang · B. Zhu (✉)
Technology Center, China Tobacco Yunnan Industrial Co., Ltd., Yunnan, China

M. Lv · B. Zhang
Beihang Yunnan Innovation Institute, Yunnan, China

X. Jia et al. (eds.), *2nd International Conference on Big Data, IoT, and Cloud Computing*, EAI/Springer Innovations in Communication and Computing,
https://doi.org/10.1007/978-3-031-93570-1_12

the release of smoke to predict changes in quality indicators. Wang Xiaohui et al. [5] proposed a method for clustering analysis of raw materials based on the chemical indicators of tobacco leaves for group compatibility. Feng Wen et al. [6] and others used the frequent pattern growth algorithm for digital maintenance of formulations and adjusted the chemical component changes according to the results. There also have formulation design methods from the perspective of cost and resource utilization rate, such as Ou Shimin et al. [7] who, based on the principle of optimal cost control, proposed the optimal formulation process for five production areas of flue-cured tobacco in the same style requirements, combining cost and inventory.

In recent years, with the accumulation of data and the improvement of computing power, research directions using data mining methods for compatibility rule analysis have emerged, such as Wang Nan et al. [8] who proposed using non-negative matrix factorization, based on similarity for single material replacement selection, to obtain the optimal solution.

Although many explorations have been made on the chemical components of tobacco formulations, the chemical components of tobacco leaves are dynamically changing, and the chemical components cannot be directly modeled with sensory qualities such as aroma, off-flavors, fineness [9], etc. Their main role is to determine the formulation and then use quality testing to measure stability. With the digitalization of tobacco formulation information, more electronic data has become possible for data mining of historical formulation maintenance methods, and intelligent recommendations through massive data analysis have been very well applied in fields such as e-commerce and market marketing [10–15].

Therefore, this study mainly did the following work:

1. Constructed a material attribute association classification method based on CBA.
2. Analyzed and applied the importance of attributes using the ID3 method.
3. Constructed a multiple linear regression model, modeled the material formulation recommendation system, and verified it.

2 Theoretical Foundation

Definition: F is the formulation set, M is the material set, both F and M are relational data sets, and a formulation in F is composed of multiple different records from M. Each material has a determined N attribute, each attribute is independent, and each attribute has multiple categories.

R is a set of historical material replacement association rules. Its association rules can be represented as $\{A \rightarrow B\}$ $(A \subset M, B \subset M, A \cap B = \varphi)$. Where A is the material to be replaced, and B is the replacement material. It indicates that A can be replaced by B, and the relationship between A and B is 1 to n.

The rule for formula material recommendation is: select a formulation from the existing formulation set, choose a material X that needs to be replaced as input. The output is a recommended set T formed by the system-recommended data group (Y, score), where (Y ⊂ M, score∈Z). The score item is the recommendation score of the material Y, and the material Y with a higher score is the top recommended material.

The following defines the rules for generating the score. Define v as the single attribute recommendation score of the material Y, α as the single attribute importance factor of the material Y. β is the reward factor, which determines the reward or penalty for the association degree between X and Y.

The product of a single attribute recommendation score and importance is the basic recommendation score for a single material and a single attribute. All attribute recommendation scores are accumulated and adjusted using the reward factor as a weight to form the score value of the material, as shown in the formula:

$$\text{Score} = \sum\nolimits_{i=1}^{n} (v_i \cdot \text{a}_i) \cdot \beta \tag{1}$$

where n is the number of material attributes, a is the single attribute importance factor. The α coefficient is based on the cumulative percentage calculation method of statistics, calculating the replacement frequency ratio of each attribute in turn, and the total percentage of the replacement frequency of all previous attributes relative to the total frequency.

v is the single attribute recommendation score of the material, and its calculation rule is: For a category of an attribute, traverse the historical replacement rules, and count the frequency with which this category is replaced by other categories. Take the top m categories with the highest frequency, normalize the frequency values, and assign a discrete score to form a score mapping table for this category being replaced by other categories as shown in Table 1.

The β factor characterizes the associative relationships among various material properties, which is derived from the analysis of frequent item sets of materials. It is defined within the Cognitive Bayesian Analysis (CBA) framework: Initially, the Apriori algorithm is utilized to mine all classification association rules. For the material replacement rule set R, the classification association rules can be represented as {A → B} (A ⊆ R, B ⊆ R, A∩B = ∅), where A and B correspond to the antecedent and consequent of the rule, respectively. In the classification association rules, B is constrained to be the categorical attribute in the classification task. Support and confidence are parameters that measure the strength of association rules; the support of a rule is the joint probability of the antecedent and consequent,

Table 1 *V* value table design

Original Category	Replacement Category	Replacement Category	*V* value
c1	c2	p1	100
c1	c3	p2	90
c1	c4	p3	80
c2	c1	p1	100
c2	c3	p2	70
c2	c4	p3	50

while the confidence is the conditional probability of the consequent given the antecedent. By setting minimum thresholds for support and confidence, appropriate rules can be generated.

2.1 Support and Confidence

The support and confidence are defined as follows:

$$\text{Sup}\,(A \to B) = P\,(A \cup B)\,(2)\text{Sup}\,(A \to B) = P\,(A \cup B) \tag{2}$$

$$\begin{aligned} \text{Conf}\,(A \to B) = P\,(A|B) = P\,(A \cup B)\,P(A)(3)\text{Conf}\,(A \to B) \\ = P\,(A|B) = P(A)P\,(A \cup B) \end{aligned} \tag{3}$$

For records that meet the confidence condition, the frequency of confidence is recorded, and after normalization, it is stored in the reward factor table for later calculation use.

The overall model process design is given in Fig. 1.

3 Model Construction

3.1 Data Preparation

Taking the 27 brands of H Group from 2016 to 2021 as the research subjects, there are 942 historical formulation records and 7532 material replacement records. Based

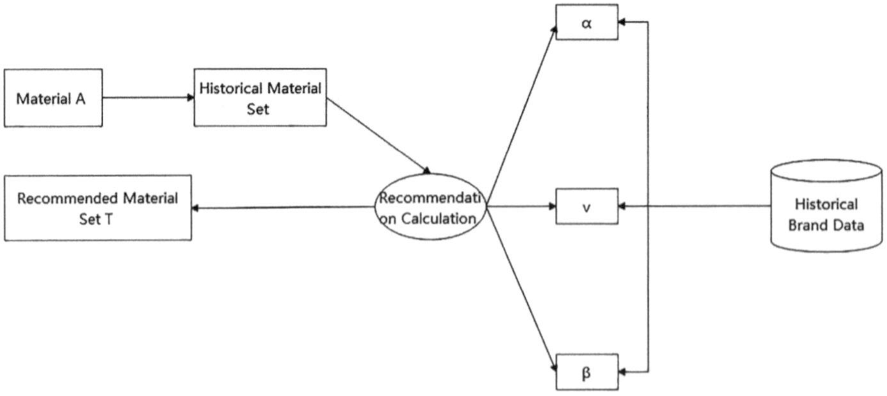

Fig. 1 Model Processing Flow

Table 2 Common Attributes and Classifications

Parameter	Values
Year	23 categories
Grade	55 categories
Major grade	17 categories
Variety	14 categories
Third-level origin	89 categories
Part	7 categories

Table 3 Attribute Information Gain Results

Attribute	Information Gain
Year	0.2879
Grade	0.1943
Variety	0.1528
Major grade	0.1277
Third-level origin	0.0551
Part	0.0412

on prior knowledge, six commonly used material attributes were selected, as shown in Table 2.

3.2 *Material Attribute Importance Coefficient*

The coefficient a is an item that describes the importance of material attributes. The analysis of attribute importance mainly focuses on the replacement priority of single material attributes [16]. Material attributes consist of year, grade, major grade, variety, third-level origin, and part, and the order of each attribute must be fully considered in material replacement.

The importance of attributes is based on decision tree analysis. The decision tree split is based on the calculation of information gain rate of each attribute by information entropy. The root node with the highest information gain has the highest value and the greatest importance. Therefore, the process from the root node split to the leaf node can be used to analyze the value and importance of attributes, with higher value attributes being prioritized for split. This study uses the ID3 algorithm and also refers to many research papers on attribute importance based on decision trees [17, 18] as the theoretical support for this method, visualizing the decision tree node split process.

After calculation, the corresponding attribute information gain is given in Table 3.

It can be seen that the importance of attributes is ranked from high to low as: Year > Grade > Variety > Major Grade > Third-level Origin > Part.

According to historical record queries, the attribute replacement frequency is given in Table 4.

Table 4 Attribute Replacement Frequency Statistics

Attribute	Replacement Frequency	Replacement Ratio
Part	891	0.051628
Major grade	1160	0.067215
Variety	1875	0.108645
Third-level origin	3556	0.206049
Grade	4071	0.235891
Year	5705	0.330571
Total	17,258	1

Table 5 Alpha Coefficient Assignment

Attribute	Year	Grade	Variety	Major Grade	Third-Level Origin	Part
α coefficient	1.0	0.67	0.43	0.23	0.12	0.05

Using the cumulative percentage of attribute replacement frequency as the value of the $a\alpha$ coefficient. According to the principle of cumulative percentage calculation, first sort the attributes from low to high importance, and then calculate the cumulative percentage for Year, Grade, Variety, Major Grade, Third-level Origin, and Part, and convert it into a decimal value as the a coefficient (Table 5).

3.3 *Material Attribute Replacement Score V*

In the recommendation algorithm, the score v describes the replacement principles of material attributes such as year, part, variety, major grade, grade, and third-level origin. By statistically analyzing the replacement frequency of each attribute of historical materials, the corresponding replacement scores for material attributes are obtained. Material attributes are divided into year, part, variety, major grade, grade, and third-level origin, for example, the year attribute has 23 values from 1998 to 2021, the part attribute has 7 values such as upper, middle, upper-middle, middle-lower, lower, whole, and other, and other attributes follow suit. The replacement frequency for each attribute value must be counted, and the top five corresponding attribute values are scored. Each attribute value is calculated to obtain the corresponding v value, and all v values are stored in the attribute value scoring table. When calculating the score, you only need to look up the table to get the corresponding v value (Fig. 2).

The design is based on the highest score being 100 with a score difference of 10. For the year attribute, the v value scores have the following corresponding relationships (Table 6).

In the preferred recommendation algorithm, the v value reflects the replacement pattern of material attributes. Different attribute items are assigned different v values according to the pattern, which allows for more concrete attention to the

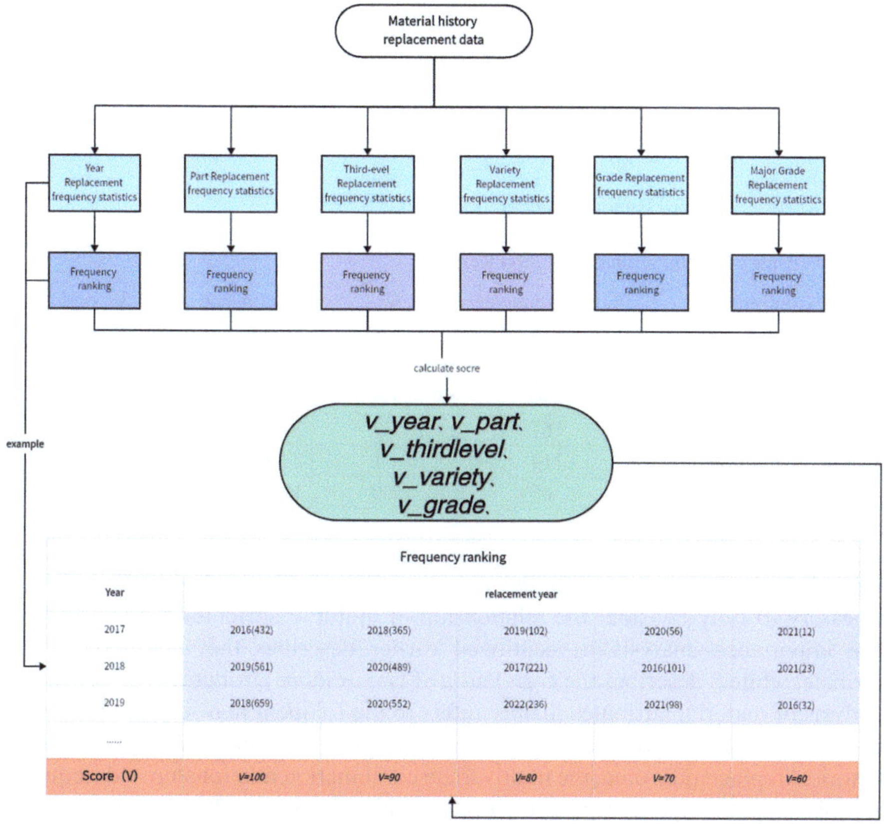

Fig. 2 Process for Material Attribute Replacement Score v

Table 6 Year Attribute *v* Value Scoring Table

Input-Recommended	2016	2017	2018	2019	2020	2021
2017	100	0	90	80	70	60
2018	70	80	0	100	90	60
2019	70	60	100	0	90	80

attribute details of recommended materials and recommended materials in the score recommendation calculation.

3.4 Material Attribute Attention Reward Factor β

In the preferred recommendation algorithm, the coefficient β describes the correlation between multiple material attributes. In formula recommendation, it is

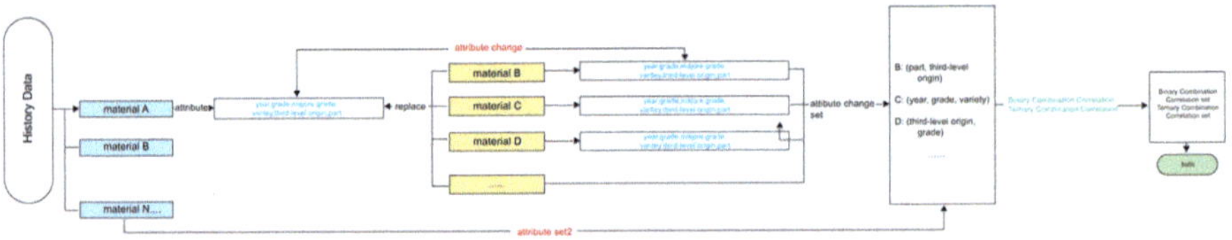

Fig. 3 Attention Reward Factor β Process

Table 7 Binary Combination Correlation Results

Binary Combination	Frequency	β	Confidence	Support
(Year, grade)	646	1.5	59.9%	35.67%
(Year, third-level origin)	589	1.4	52.56%	32.52%
(Third-level origin, grade)	230	1.3	32.64%	12.7%
(Year, variety)	167	1.2	31.3%	9.22%
(Variety, third-level origin)	114	1.1	23.16%	6.2%
Others	65	1.0	12.7%	3.58%

not only necessary to consider the priority of one attribute, but sometimes it is necessary to fully consider the relationship of multiple attributes for replacement. The importance analysis α mentioned earlier describes the importance of an attribute, while β describes the correlation of two or more attributes. The correlation analysis of material attributes mainly refers to the frequent itemset of collaborative filtering algorithms. By counting the most frequent attribute combinations in the historical replacement data, the multivariate combination relationship and frequency of replacement material attributes are statistically analyzed. The more frequently a combination appears, the closer the relationship and the higher the attention. For details, see Fig. 3.

After the correlation analysis of material attributes, the following corresponding combinations, frequencies, and reward factors β are obtained as shown in Table 7:

In the recommendation system, confidence is the conditional probability that the elements of the combination occur simultaneously, and support is the probability of the combination occurring. When making recommendations, both confidence and support should be combined to balance the recommendation conditions, where confidence is calculated as the average value of the combination.

From the above relationship combinations, it can be seen that the most frequent replacement relationship in binary attribute combinations is Year and Grade, and in the selection of replacement materials, Year and Grade are more focused on attention. In ternary attribute combinations as shown in Table 8, the most frequent replacement relationship is Year, Third-Level Origin, and Grade, and in the selection of replacement materials, more attention is paid to the attribute combination of Year, Third-Level Origin, and Grade. According to the above attribute combination frequency, the optimal combination reward β is obtained, and the final binary and ternary combination reward scores are stored in the attention query table. When

Table 8 Ternary Combination Correlation Results

Ternary Combination	Frequency	β	Confidence	Support
(Year, third-level origin, grade)	678	1.5	64.57%	50.63%
(Year, third-level origin, variety)	217	1.4	63.12%	16.2%
(Year, variety, grade)	161	1.3	62.26%	12.02%
(Part, major grade, grade)	123	1.2	52.17%	9.18%
(Third-level origin, part, grade)	118	1.1	38.48%	8.81%
Others	42	1.0	27.2%	3.13%

Table 9 Single Material Recommendation Verification

Replaced Material	Historical Replacement Set (R)	Historical Replacement Set (R)	B∩R	Recommendation Hit Rate
A	B, M, D, E, F	B, C, D, E, F	B, D, E, F	80%

calculating the score, only the attribute changes of the two materials need to be analyzed, and the corresponding reward score can be found.

4 Results and Analysis

4.1 Model Verification

By comparing the material recommendations from the model with the replacement history, the algorithm's recommendations are verified against historical replacements. The specific verification process is as follows:

For material A, the model recommends results for materials B, C, D, E, and F. Material A has been replaced by materials B, D, E, F, and M in historical replacements, among which material C has never appeared in history, then the model's applicability is 80% (Table 9). Since the number of replacement data records varies for different materials, the experiment only takes five random historical replacement records for each material as verification data.

Randomly extract 5 material numbers from the system: 148237075, 148,556,701, 153,668,972, 146,933,775, and 152,887,965 for verification, and the results are given in Table 10:

Using five materials for verification, there are a total of 19 historical replacement records for the five materials, and 17 of the model's recommended results are consistent with historical replacements, so the applicability verification is 89.4%.

All data between 5 years is calculated to generate a material recommendation table of 24,680, and the test data uses the historical replacement data of 2023 for cross comparison. The applicability verification result is 92%. The experiment

Table 10 5 Material Recommendation Verification

Replaced Material (A)	Recommended Set (B)	Historical Replacement Set (R)	B∩R record
148237075	149321075 149171075 148249075 150214075 149668075	148249075 149171075 149321075 149668075 150214075	5
148556701	153668972 143215487 148996871 149171075 14989897	143215487 14989897 149171075 N/A N/A	3
153668972	143215487 149171075 149668075 152669784 14423654	149171075 14423654 149668075 N/A N/A	3
146933775	148995671 149171075 149171075 14989897 149321075	149171075 149321075 148995652 148995671 N/A	4
152887965	153668972 157998264 158223169 149775698 156785423	143215789 157998264 158223169 156785423 N/A	2

proves that the recommendation results of the algorithm model are consistent with historical replacement experience and can replace manual work in actual production.

4.2 Model Analysis

The recommendation algorithm is essentially a linear equation about v year, v grade, v variety, v major grade, v third-level origin, v part, and X, that is, the x coefficient can be modeled and learned through machine learning algorithms as follows:

$$\text{Score}=\boldsymbol{X_0\beta v}_{year}+\boldsymbol{X_1\beta v}_{grade}+\boldsymbol{X_2\beta v}_{variety}+\boldsymbol{X_3\beta v}_{\text{level}}+\boldsymbol{X_4\beta v}_{\text{thirdlevel}}+\boldsymbol{X_5\beta v}_{\text{part}} \quad (4)$$

In the formula, v values and β are known, so if the Score is also known, the formula can be used as the objective function to calculate the loss and optimize the calculation error of the algorithm. Therefore, using real data to correct the algorithm's results is the overall goal of optimization, and the true value of Score is the key to optimization. Real data needs to collect user feedback data, and the recommendation algorithm is corrected by user feedback data to optimize the deviation of Score, making the model's recommendation results more accurate.

5 Conclusion

By data mining the historical data of 27 brands of Company H over the past 6 years, a corresponding formula recommendation algorithm was established and verified through association rules. The results show that the association rule recommendation method is highly applicable for formula maintenance, with an accuracy rate of 92%. This calculation method is very beneficial for strengthening the weight of inventory materials, improving the utilization rate of materials, and enhancing the work efficiency of formula experts, and reducing the work pressure of evaluation personnel. Currently, the algorithm does not consider the recommendation of priorities, and there is no mature technology for newly developed formulas. Therefore, if historical data and chemical indicators are combined, using a fusion model can create greater value in product digital design.

Acknowledgments Technical Center of China Tobacco Yunnan Co., LTD Intelligent recommendation of leaf group recipe Visualization based on Big data machine learning.

References

1. Nan, W.: Research on Data-Driven Tobacco Formula Maintenance Methods. Northeastern University (2019)
2. Chao, D., Xue-xia, L., Zhi, C., et al.: Six portrait systems based on big data label system of tobacco consumption. Chinese Tobacco Sci. **30**, 1 (2024)
3. Jianping, X.: Situation and future: Prospect of tobacco technology development. Chinese Tobacco Sci. **2017**, 23 (2017)
4. Jianmin, W., Xuewen, G., Xiao, L., Wei, P.G., Tianbing, Z.: Establishment of tobacco formula and tobacco smoke prediction model. Tobacco Sci. **6**, 5 (2006)
5. Xiaohui, W., Bin, Y., Guozhi, T., Liyue, D.: Tobacco formula tobacco grouping method based on chemical characteristics of tobacco raw materials. Tobacco Sci. **3**, 5–7, 22 (2013)
6. Ou Shimin, M., Dongsheng, P.W., Wujun, F., Meiling, S., Gaosong, L., Qiang, W., Deping, J., Peng, H.: Research on the Design of Tobacco Formula Module for mellow and delicate style of flue-cured tobacco. Agric. Technol. **40**(6), 1–5 (2020)
7. Wen, F., Jiba, C., Ting, L., Haitao, L., Zhimin, Y.: Changes in chemical components in tobacco formula digital maintenance engineering. Food Industry. **40**(11), 199–202 (2019)
8. Nan, W., Xingang, L., Zhongliang, Z., Yang, Y., Jiafu, T., Jianguo, T.: Tobacco formula maintenance method based on non-negative matrix factorization. Tobacco Sci. **52**(8), 67–76 (2019)
9. Ziying, Z., Weimin, G., Wentao, X., et al.: Selection of key appearance feature indicators for sensory quality of flue-cured tobacco based on partial least squares regression. J. Light Indust. **38**(4), 90–97 (2023)

10. Qi, H., Dingju, Z., Huilin, W., Lihong, W.: A review of research on intelligent recommendation systems. Comput. Syst. Appl. **31**(4), 47–58 (2022)
11. Bin, Z.: Research on intelligent recommendation system for medical clinical pathways based on collaborative filtering. Modern Electr. Technol. **46**(4), 75–78 (2023)
12. Mengzhu, J., Guang, Z.Y., Fei, L., Donglan, F.: Research on optimization of supply chain decision making under the application of big data intelligent recommendation technology. Sci. Technol. Manag. Res. **42**(9), 162–167 (2022)
13. Wei, T., Dong, L.: Research on intelligent recommendation algorithm for archives based on collaborative filtering. Lantai World. **2**, 84–88 (2024)
14. Junyi, Z., Fuzhen, Z., Xiang, A., Qing, H., Huiqin, J., Ling, M.: A review of collaborative filtering recommendation systems. J. Inf. Secur. **6**(5), 17–34 (2021)
15. Xiaohan, Y., Guosheng, H., Xiehua, Z., Zihao, Y.: Collaborative filtering algorithm based on co-training and boosting. Comput. Appl. **43**(10), 3136–3141 (2023)
16. Lin, W., Pingcong, Z., Tiejun, Y., Yuhan, G., Ping, Z., Qinghua, W.: Research on the analysis model of tobacco formula module value measurement. Hubei Agric. Sci. (2024)
17. Yahui, C., Jihua, Y.: A personalized agricultural product Mobile information service system based on decision tree classification. J. Jiangxi Normal Univ. (Natural Science Edition). **2**, 145–148 (2016)
18. Chungui, L., Qingneng, Y., Meng, W., Xianhua, R.: A decision tree algorithm based on attribute frequency division. J. Guangxi Univ. Technol. **4**, 1–4 (2007)

An Efficient Video Salient Object Detection Method for IoT

Qinmen Chen, Jiale Ru, Caisheng Liu, and Dan Xu

1 Introduction

Video salient object detection aims to segment the most visually prominent regions from a set of video frames. Unlike saliency detection in static images, video saliency detection poses additional challenges as it must consider spatial and temporal information and the interaction between these two feature spaces [1, 2]. Video saliency detection is a research hotspot in the fields of computer vision and the Internet of Things (IoT), with broad applications in video editing, video caption generation, autonomous driving, smart homes, drones, and so on [3–5]. Applications in the IoT domain demand higher standards from video saliency detection algorithms. Firstly, the algorithm must demonstrate robust performance, capable of handling various image scenarios in open environments, such as low lighting, occlusion, motion blur, and noise interference. Secondly, it should be adaptable to the limited computational power of edge devices while meeting the real-time requirements of edge-end applications.

Currently, the main research methods for video saliency object detection can be classified into three categories: methods based on 3D convolution [1, 2]; methods based on Convolutional Long Short-Term Memory networks (ConvLSTM) [6, 7]; and approaches based on optical flow maps [8]. Saliency detection algorithms based on 3D convolution encode or decode spatial-temporal information simultaneously. For example, Chen et al. [2] proposed a Pyramid Dilated 3D Convolution (PD3C) module, which generates temporal features by utilizing contextual information.

Q. Chen
Nanjing Marine Radar Research Institute, Nanjing, China

J. Ru · C. Liu · D. Xu (✉)
Jiangsu University of Science and Technology, Zhenjiang, China

X. Jia et al. (eds.), *2nd International Conference on Big Data, IoT, and Cloud Computing*, EAI/Springer Innovations in Communication and Computing,
https://doi.org/10.1007/978-3-031-93570-1_13

ConvLSTM primarily utilizes the temporal perception ability of recurrent networks to extract temporal information from video frames. Inspired by human attention transfer behavior, Fan et al. [6] proposed a saliency transfer-aware ConvLSTM module to extract dynamic saliency information, while Zhao et al. [7] introduced a bidirectional ConvLSTM model for long-term time context modeling. However, due to insufficient spatiotemporal interaction of ConvLSTM models, the algorithms often produce blurry saliency maps. Video saliency detection algorithms based on optical flow maps primarily utilize spatial features from RGB images and temporal features from optical flow maps. For instance, Liu et al. [8] constructed a symmetric dual-branch network that adaptively learns spatial features from RGB images and temporal features from optical flow maps using a dynamic weighting generator. Video saliency detection algorithms based on optical flow heavily depend on the quality of the optical flow maps; low-quality maps degrade detection results, while high-quality maps increase algorithmic time costs.

To mitigate the trade-off between algorithm accuracy and time consumption, some scholars have proposed fast video saliency detection algorithms. Huang et al. [9] introduced a lightweight video saliency detection method using ShuffleNet as the backbone network for feature extraction. However, ShuffleNet's feature extraction capability is relatively weak, which affects the detection accuracy of the algorithm. Zhao et al. [10] employed 3D convolution for unified spatiotemporal modeling and alleviated temporal degradation issues using a random interaction mechanism. However, 3D convolution models have a large number of parameters, posing challenges for optimization on edge devices.

To address the application needs of video saliency detection in the Internet of Things (IoT), this paper proposes a novel and efficient video saliency object detection algorithm. The algorithm enhances accuracy and reduces computational time by focusing on key aspects such as backbone networks, temporal feature extraction, and spatiotemporal information fusion. The primary contributions of this algorithm are as follows:

1. We propose a lightweight spatiotemporal interaction framework that adheres to an encoding-decoding structure. Each encoding layer is connected to its corresponding decoding layer via an attention mechanism. In the decoding phase, we design Inter-Frame Self-Attention (IFSA) modules and Attentional ConvGRU modules at different decoding layers to rapidly extract precise temporal information. This framework allows the network loss to propagate from the temporal (decoding phase) to the spatial domain (encoding phase), addressing performance bottlenecks caused by insufficient spatiotemporal interaction in existing algorithms.
2. IFSA modules are integrated into the deep decoding stages to simultaneously extract spatiotemporal information. These modules compute similarity matrices between the current frame and others, enabling precise localization of prominent regions. Furthermore, by comparing the relationships between the current frame and other frames in the video set, long-term motion trends of the object can be captured.

3. We introduce the AttConvGRU module in the shallow layers of the decoding network. This module comprises a lightweight ConvGRU combined with channel-spatial attention. The lightweight ConvGRU is used for extracting temporal information, while the channel-spatial attention helps minimize noise introduced between adjacent frames.
4. We compare our proposed algorithm with mainstream video saliency detection algorithms and fast video saliency detection algorithms on multiple public datasets. The results demonstrate that our algorithm achieves state-of-the-art performance in both accuracy and efficiency.

2 The Proposed Method

2.1 *Overview of the Framework*

As shown in Fig. 1, the proposed method consists of two parts: encoding and decoding. The encoding phase employs a transformer network [11] as the backbone for spatial feature extraction, while the decoding phase uses fully convolutional operations as the basic decoding module. To fully utilize multi-scale information in the encoding phase, attention modules are employed to connect each encoding and decoding layer. In the deep layers of the decoding phase, inter-frame self-attention (IFSA) mechanisms are used for spatial localization of salient objects and extraction of inter-frame temporal information. In the shallow layers of the decoding phase, ConvGRU modules with attention are used to further refine displacement information of salient objects. The utilization of diverse temporal feature extraction modules, tailored to distinct layers in the decoding phase, aids in the rapid and precise localization of salient objects within video frames.

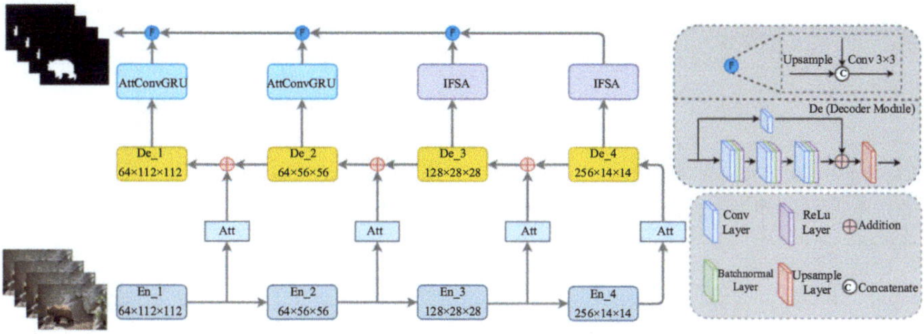

Fig. 1 Overall architecture of the proposed method, where IFSA represents the inter-frame self-attention module, AttConvGRU represents the attention ConvGRU module, and Att denotes attention mechanism modules

2.2 Inter-Frame Self-Attention Module (IFSA)

The decoding network layers, from deep to shallow, exhibit a gradual refinement in the process of object localization. To elevate localization precision, we have incorporated inter-frame attention mechanism modules within the deeper layers of the decoding network. IFSA module consists of two steps: firstly, using self-attention mechanisms on each video frame for saliency object localization; secondly, employing inter-frame cooperative attention mechanisms to enhance the collaboration between frames, thereby achieving precise saliency object localization. Simultaneously, pixel-wise comparisons between frames capture object motion trends, thereby obtaining inter-frame temporal information, thus achieving unified temporal and spatial information. Note that we solely rely on pixel-wise multiplication operations in computing frame similarities, eliminating time-consuming convolution operations and minimizing computational overhead.

The details of the IFSA module are illustrated in Fig. 2. This figure primarily demonstrates the process where frame f_1 undergoes the IFSA module to obtain $f_1^{'}$, with similar computations applied to the other frames f_2, f_3, and f_4 to obtain $f_2^{'}, f_3^{'}$, and $f_4^{'}$. The IFSA module comprises the following four steps.

Step 1: convolutional layers with kernel size 3 × 3 are applied on the input video frames $\mathrm{f_t} \in \mathbb{R}^{1 \times C \times H \times W} : \mathrm{t} = 1, 2, 3, 4$ to acquire Query $\mathrm{Q_t} \in \mathbb{R}^{1 \times C \times H \times W}$, Key $\mathrm{K_t} \in \mathbb{R}^{1 \times C \times H \times W}$, and Value $\mathrm{V_t} \in \mathbb{R}^{1 \times C \times H \times W}$. The step can be formulated as follows:

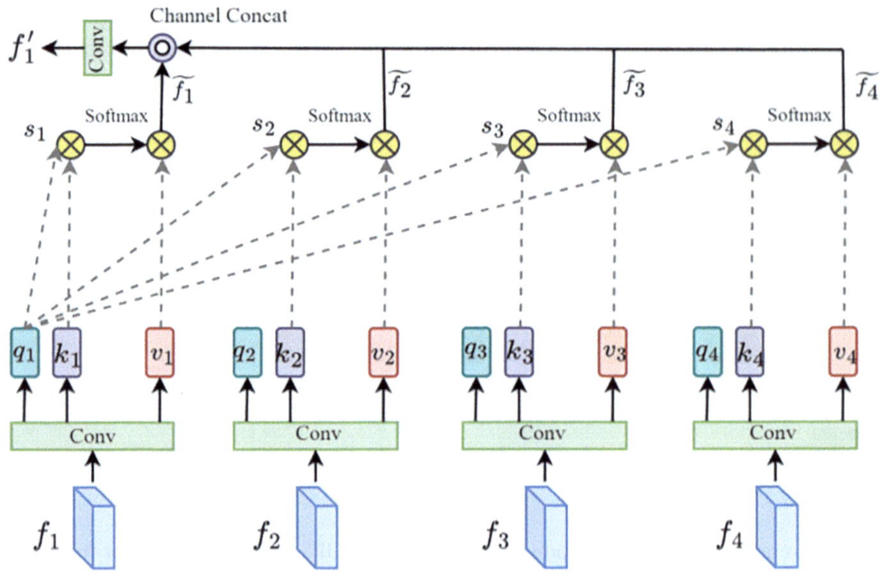

Fig. 2 The inter-frame self-attention module

$$Q_t = \theta(f_t), K_t = \varphi(f_t), V_t = \rho(f_t) \tag{1}$$

where θ, φ, and ρ are the convolutional layers.

Step 2: compute the self-attention for frame f_1. Calculate the similarity between the query q_1 and key k_1 of frame f_1, obtaining the similarity matrix $s_1 = q_1 \otimes k_1$, where $\otimes$ denotes matrix dot product. Then, transform the similarity matrix into a weight matrix using the softmax function to weight v_1, yielding feature map $\widetilde{f_1}$ containing salient object positional information.

Step 3: compute inter-frame cooperative attention. Calculate the similarity between the query q_1 and key k_t: $t = 2,3,4$, obtaining the similarity matrix $s_t = q_1 \otimes k_t$. Transform the similarity matrix s_t into weight matrices using the softmax function and perform element-wise multiplication with v_t, yielding feature map $\widetilde{f_t}$. The feature map $\widetilde{f_t}$ captures motion trends between video frames and displacement of salient objects across frames.

Step 4: the positional information of the salient object, $\widetilde{f_1}$, and the displacement information between different frames, $\widetilde{f_t}$, are concatenated in the channel dimension. Then, a unified spatiotemporal feature map is obtained through a convolutional layer. The calculation formula for this step is as follows:

$$\mathrm{f}_1' = \mathrm{Conv}_{1\times 1}\left(\mathrm{cat}\left(\widetilde{\mathrm{f}_1}, \widetilde{\mathrm{f}_2}, \widetilde{\mathrm{f}_3}, \widetilde{\mathrm{f}_4}\right)\right) \tag{2}$$

where cat represents concatenation in the channel dimension.

2.3 ConvGRU with Channel-Spatial Attention (AttConvGRU)

The ConvLSTM is widely used for extracting temporal information in video saliency detection. However, the following characteristics have limited its use in fast video saliency detection: firstly, the bidirectional propagation network leads to complexity and a large number of parameters; secondly, extracting long-term temporal information introduces inter-frame noise, which degrades detection performance; thirdly, the network fails to effectively utilize spatial multi-scale information, resulting in performance bottlenecks. To address these issues, ConvGRU is employed as a replacement for ConvLSTM to simplify the temporal information extraction module while maintaining accuracy. Subsequently, a channel-spatial attention mechanism is designed to filter inter-frame noise after ConvGRU, thereby effectively extracting temporal features. Finally, by integrating the attConvGRU module at different decoding stages, spatial multi-scale information is effectively harnessed.

The detailed diagram of the ConvGRU module with the attention mechanism is shown in Fig. 3. The module can be divided into two parts: one part is the

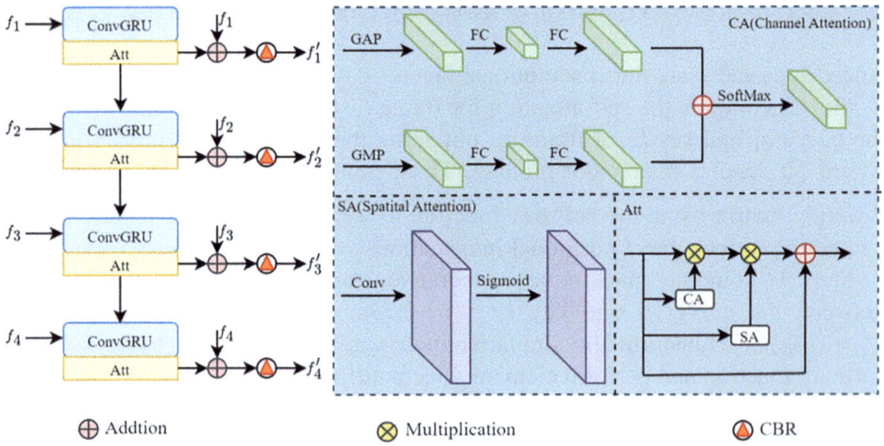

Fig. 3 Detailed structure of the proposed AttConvGRU module

ConvGRU module, and the other part is the attention mechanism module. For a set of input video frames, the attention ConvGRU module can obtain feature maps which contain temporal information. This process can be represented by the following formula:

$$f_t' = CBR\left(f_t + Att\left(ConvGRU\left(f_t, f_{t-1}\right)\right)\right) \tag{3}$$

where *CBR* denotes the combination of Conv, batchnormal, and ReLu. *Att* represents a channel-spatial attention module that consists of two parts: channel attention mechanism (CA) and spatial attention mechanism (SA). The channel attention utilizes global average pooling (GAP) and global maximum pooling (GMP) for dimension compression, which can be expressed as follows:

$$\mathrm{CA_{gap}} = \mathrm{FC_1}\left(\mathrm{FC_2}\left(\mathrm{GAP}(f)\right)\right) \tag{4}$$

$$\mathrm{CA_{gmp}} = \mathrm{FC_1}\left(\mathrm{FC_2}\left(\mathrm{GMP}(f)\right)\right) \tag{5}$$

$$\mathrm{CA} = \mathrm{Softmax}\left(\mathrm{CA_{gap}} + \mathrm{CA_{gmp}}\right) \tag{6}$$

Here, $\mathrm{FC_1}$ and $\mathrm{FC_2}$ represent fully connected layers that map the feature map dimension to [C/2, 1, 1] and [C, 1, 1], respectively.

The spatial attention can be expressed with the convolution and sigmoid operations with formula:

$$\mathrm{SA} = \mathrm{Sigmoid}\left(\mathrm{Conv}(f)\right) \tag{7}$$

where Conv represents the convolution layer of size 1×1, transforming the feature map dimension from [C, H, W] to [1, H, W].

Finally, the output of the attention module is provided with the formula:

$$Att = f + f \bigotimes SA \bigotimes CA \tag{8}$$

where $\bigotimes$ represents pixel-wise multiplication.

3 Experiments

3.1 Implementation and Experimental Setup

We first pre-train the backbone network, Shunted-Transformer, on the static saliency detection dataset, DUTS-TR [12], followed by end-to-end training on the video saliency detection datasets DAVIS [13] and DAVSOD [6]. Finally, we evaluate our method on five datasets: DAVIS, SegV2 [14], ViSal [15], VOS [16], and DAVSOD. Input images are resized to 320×320 and undergo random flips, crops, rotations, and color enhancements to prevent overfitting. We employ the AdamW optimization algorithm with cosine decay learning rate scheduling. During the pre-training and end-to-end training phases, the learning rate decreases from 1e-4 to 1e-5, and then from 1e-5 to 1e-6 over 20 iterations each. The batch size is set to 1, with each batch containing four video frames. All training and testing experiments are conducted on an NVIDIA GeForce GTX Titan XP GPU.

3.2 Experiment Results and Analysis

We conducted quantitative, qualitative, and time consumption comparison with 11 state-of-the-art video saliency detection algorithms. We use three metrics to evaluate our approach: mean absolute error (MAE), F-measure (F_β), and structural similarity (S_λ) [6]. Additionally, frames per second (FPS) is used to assess the time efficiency of the algorithms. The comparison results demonstrate that our algorithm achieves favorable performance in both detection precision and time consumption, achieving the purpose of efficient video saliency detection.

Quantitative Evaluation

Table 1 presents the comparative results of our method with other 16 algorithms on five commonly used video saliency detection datasets. It can be observed that our method achieves the best results compared to previous algorithms on the SegV2 and

Table 1 Quantitative comparisons of three evaluation metrics on five VSOD datasets. "↑" and "↓" indicate that larger or smaller is better. "**" represents that the algorithm doesn't provide the experiment results. The top two results are marked in boldface and underscore

Methods		RCRNet	SSAV	TENet	STVS	STANet	MQP	DCFNet	DSNet	TCRNet	SCANet	PSNet	Ours
Metrics		[17]	[6]	[18]	[19]	[20]	[21]	[22]	[8]	[23]	[1]	[24]	
DAVIS	$S_\lambda\uparrow$	0.864	0.893	0.905	0.892	0.892	0.916	0.914	0.914	0.916	0.902	**0.919**	0.916
	$F_\beta\uparrow$	0.828	0.861	0.881	0.862	0.883	0.904	0.900	0.891	0.898	0.881	**0.907**	0.901
	MAE↓	0.036	0.028	0.017	0.023	0.025	0.018	0.016	0.018	0.016	0.021	0.016	**0.014**
SegV2	$S_\lambda\uparrow$	0.899	0.851	0.868	0.891	0.872	0.882	0.883	0.875	0.904	0.906	0.889	**0.934**
	$F_\beta\uparrow$	0.869	0.801	0.810	0.860	0.769	0.841	0.839	0.832	0.865	0.890	0.852	**0.930**
	MAE↓	0.013	0.023	0.025	0.017	0.031	0.018	0.015	0.028	0.014	0.026	0.016	**0.009**
ViSal	$S_\lambda\uparrow$	**0.959**	0.943	0.949	0.952	0.910	0.942	0.952	0.949	0.946	0.954	0.954	0.954
	$F_\beta\uparrow$	0.951	0.939	0.949	0.952	0.893	0.939	0.953	0.950	0.943	0.955	0.955	**0.960**
	MAE↓	0.013	0.020	0.012	0.013	0.021	0.016	0.010	0.013	0.013	0.011	0.012	**0.009**
VOS	$S_\lambda\uparrow$	**	0.819	0.845	0.850	0.804	0.828	0.846	0.855	0.867	**0.872**	**	0.840
	$F_\beta\uparrow$	**	0.742	0.781	0.791	0.701	0.768	0.791	0.801	0.823	0.828	**	**0.830**
	MAE↓	**	0.073	0.052	0.058	0.062	0.069	0.060	0.060	0.049	**0.048**	**	0.050
DAVSOD	$S_\lambda\uparrow$	0.780	0.724	0.779	0.746	0.706	0.770	0.741	0.729	**	**0.801**	0.765	0.772
	$F_\beta\uparrow$	0.696	0.603	0.697	0.651	0.549	0.703	0.660	0.627	**	**0.731**	0.678	0.703
	MAE↓	0.062	0.092	0.070	0.086	0.107	0.075	0.074	0.077	**	0.064	0.074	**0.061**

ViSal datasets. In terms of the MAE evaluation metric on the DAVIS and DAVSOD datasets, our algorithm also performs the best. Our algorithm did not achieve optimal results on the VOS and DAVSOD datasets in both evaluation metrics S_λ and F_β, primarily because our inter-frame self-attention module focuses more on changes between adjacent and nearby frames. The VOS dataset includes a ground truth label only every 15 frames, creating significant temporal gaps that challenge our method. The DAVSOD dataset contains continuously changing salient objects, which also present challenges for our algorithm.

Qualitative Evaluation

We conducted a comparative analysis of our proposed method against eight advanced video saliency detection algorithms across diverse video scenes, as illustrated in Fig. 4. Our approach demonstrates robust capabilities in accurately identifying salient objects across challenging scenarios, including cluttered backgrounds and small objects (first and second rows), large objects (third and fourth rows), objects at image edges (fifth and sixth rows), complex foregrounds (seventh and eighth rows), low contrast scenes, and scenarios with multiple objects (ninth and tenth rows). Leveraging an inter-frame self-attention mechanism with pixel-wise comparison at the deep layers of the decoding stage facilitates the approximate localization of salient objects through collaborative analysis of input frames. Additionally, integration of the attention ConvGRU module at the shallow decoding layers enhances detailed information representation, resulting in superior detection performance across varied challenging scenarios.

Time Consumption Evaluation

Most existing video saliency detection algorithms are based on 3D convolutions, ConvLSTM, and optical flow methods, which often struggle to balance the trade-off between detection accuracy and speed. In contrast, our method leverages a lightweight backbone network and a temporal information extraction module, achieving a high detection rate while maintaining accuracy. A comparison of the frame rates between our approach and other methods is shown in Table 2.

4 Conclusions

This paper presents an efficient video saliency object detection algorithm composed of two key modules. The first module is an inter-frame self-attention mechanism designed to precisely localize salient objects and extract deep temporal features from the network. By leveraging the mutual interactions between frames, this module enhances object localization and temporal information extraction. The

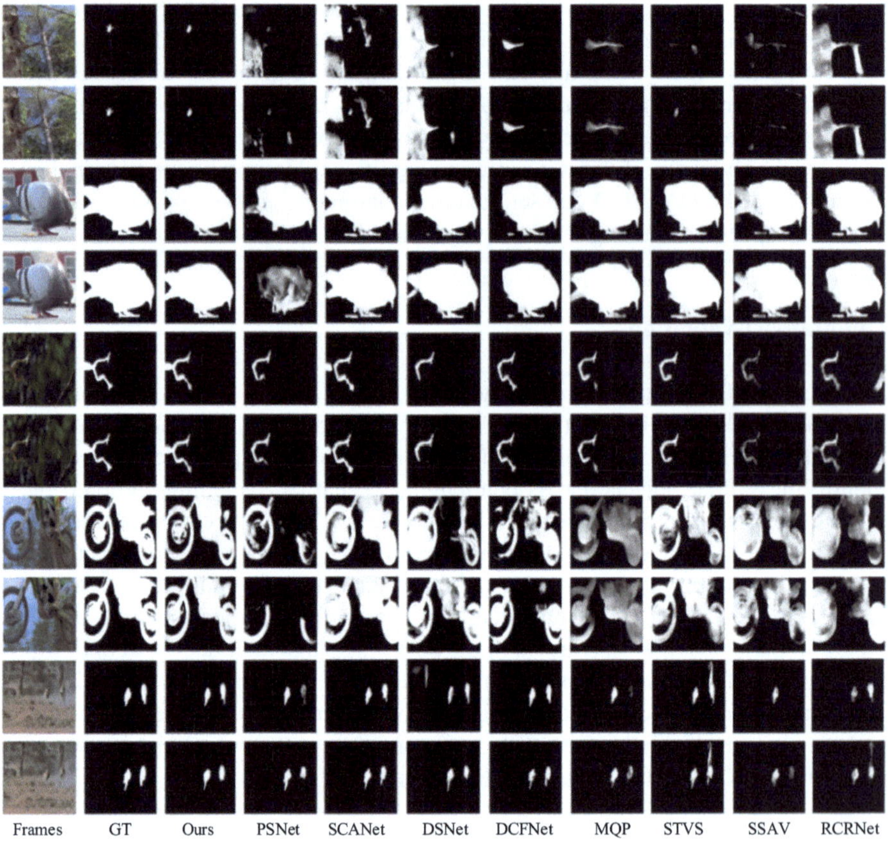

Fig. 4 Qualitative comparisons of our approach with eight existing state-of-the-art VSOD methods

second module, Attention ConvGRU, focuses on extracting shallow temporal feature information. By incorporating attention mechanisms between ConvGRU modules, this approach aims to reduce temporal consumption and filter out noise from previous frames, thereby improving detection accuracy. Compared to other methods, our approach better balances the trade-off between temporal consumption and detection accuracy. In future work, we will further investigate the application of the proposed method on edge devices.

Table 2 The FPS comparison between our method and eight other algorithms

Methods	SCANet	DCFNet	MQP	STANet	STVS	TENet	SSAV	RCRNet	Ours
Based on	3D conv	Context	Optical flow	ConvLSTM	3D conv	–	ConvLSTM	ConvGRU	ConvGRU
GPU	Titan Xp	2080Ti	2080Ti	V100	1080Ti	1080Ti	Titan Xp	1080Ti	Titan Xp
FPS	41	28	33	17	50	17	20	27	57

References

1. Chen, T., Xiao, J., Hu, X., Zhang, G., Wang, S.: Spatiotemporal context-aware network for video salient object detection. Neural Comput. & Applic. **34**(19), 16861–16877 (2022). https://doi.org/10.1007/s00521-022-07330-1
2. Chen, Q., Liu, Z., Zhang, Y., Fu, K., Zhao, Q., Du, H.: RGB-D salient object detection via 3D convolutional neural networks. In: Proceedings of the AAAI Conference on Artificial Intelligence, vol. 35, pp. 1063–1071 (2021, May). https://doi.org/10.1609/aaai.v35i2.16191
3. Singh, H., Verma, M., Cheruku, R.: DSNet: efficient lightweight model for video salient object detection for IoT and WoT applications. In: Companion Proceedings of the ACM Web Conference 2023, pp. 1286–1295 (2023, April). https://doi.org/10.1145/3543873.3587592
4. Chen, Y., Huang, J., Mou, L., Jin, P., Xiong, S., Zhu, X.X.: Deep saliency smoothing hashing for drone image retrieval. IEEE Trans. Geosci. Remote Sens. **61**, 1–13 (2023). https://doi.org/10.1109/TGRS.2023.3255302
5. Shin, C., Kim, H., Lee, C.H., Lee, S.G., Yoon, S.: Edit-a-video: single video editing with object-aware consistency. In: Asian Conference on Machine Learning, pp. 1215–1230. PMLR (2024)
6. Fan, D.P., Wang, W., Cheng, M.M., Shen, J.: Shifting more attention to video salient object detection. In: Proceedings of the IEEE/CVF conference on computer vision and pattern recognition, pp. 8554–8564 (2019)
7. Zhao, W., Zhang, J., Li, L., Barnes, N., Liu, N., Han, J.: Weakly supervised video salient object detection. In: Proceedings of the IEEE/CVF Conference on Computer Vision and Pattern Recognition, pp. 16826–16835 (2021)
8. Liu, J., Wang, J., Wang, W., Su, Y.: DS-net: dynamic spatiotemporal network for video salient object detection. Digit. Signal Process. **130**, 103700 (2022). https://doi.org/10.1016/j.dsp.2022.103700
9. Huang, K., Xu, Z.: Lightweight video salient object detection via channel-shuffle enhanced multi-modal fusion network. Multimed. Tools Appl. **83**(1), 1025–1039 (2024). https://doi.org/10.1007/s11042-023-15251-x
10. Zhao, X., Liang, H., Li, P., Sun, G., Zhao, D., Liang, R., He, X.: Motion-aware memory network for fast video salient object detection. IEEE Trans. Image Process. (2024). https://doi.org/10.1109/TIP.2023.3348659
11. Ren, S., Zhou, D., He, S., Feng, J., Wang, X.: Shunted self-attention via multi-scale token aggregation. In: Proceedings of the IEEE/CVF Conference on Computer Vision and Pattern Recognition, pp. 10853–10862 (2022)
12. Wang, L., Lu, H., Wang, Y., Feng, M., Wang, D., Yin, B., Ruan, X.: Learning to detect salient objects with image-level supervision. In: Proceedings of the IEEE Conference on Computer Vision and Pattern Recognition, pp. 136–145 (2017)
13. Perazzi, F., Pont-Tuset, J., McWilliams, B., Van Gool, L., Gross, M., Sorkine-Hornung, A.: A benchmark dataset and evaluation methodology for video object segmentation. In: Proceedings of the IEEE Conference on Computer Vision and Pattern Recognition, pp. 724–732 (2016)
14. Li, F., Kim, T., Humayun, A., Tsai, D., Rehg, J.M.: Video segmentation by tracking many figure-ground segments. In: Proceedings of the IEEE International Conference on Computer Vision, pp. 2192–2199 (2013)
15. Wang, W., Shen, J., Shao, L.: Consistent video saliency using local gradient flow optimization and global refinement. IEEE Trans. Image Process. **24**(11), 4185–4196 (2015). https://doi.org/10.1109/TIP.2015.2460013
16. Li, J., Xia, C., Chen, X.: A benchmark dataset and saliency-guided stacked autoencoders for video-based salient object detection. IEEE Trans. Image Process. **27**(1), 349–364 (2017). https://doi.org/10.1109/TIP.2017.2762594
17. Su, Y., Deng, J., Sun, R., Lin, G., Wu, Q.: A unified transformer framework for group-based segmentation: co-segmentation, co-saliency detection and video salient object detection. IEEE Trans. Multimed. **26**, 313–325 (2023). https://doi.org/10.1109/TMM.2023.3264883

18. Ren, S., Han, C., Yang, X., Han, G., He, S.: Tenet: triple excitation network for video salient object detection. In: Computer vision–ECCV 2020: 16th European conference, Glasgow, UK, August 23–28, 2020, Proceedings, part V 16, pp. 212–228. Springer, Cham (2020)
19. Chen, C., Wang, G., Peng, C., Fang, Y., Zhang, D., Qin, H.: Exploring rich and efficient spatial temporal interactions for real-time video salient object detection. IEEE Trans. Image Process. **30**, 3995–4007 (2021). https://doi.org/10.1109/TIP.2021.3068644
20. Bi, H.B., Lu, D., Zhu, H.H., Yang, L.N., Guan, H.P.: STA-net: spatial-temporal attention network for video salient object detection. Appl. Intell. **51**, 3450–3459 (2021). https://doi.org/10.1007/s10489-020-01961-4
21. Chen, C., Song, J., Peng, C., Wang, G., Fang, Y.: A novel video salient object detection method via semisupervised motion quality perception. IEEE Trans. Circuits Syst. Video Technol. **32**(5), 2732–2745 (2021). https://doi.org/10.1109/TCSVT.2021.3095843
22. Zhang, M., Liu, J., Wang, Y., Piao, Y., Yao, S., Ji, W., Luo, Z.: Dynamic context-sensitive filtering network for video salient object detection. In: Proceedings of the IEEE/CVF International Conference on Computer Vision, pp. 1553–1563 (2021)
23. Huang, K., Tian, C., Su, J., Lin, J.C.W.: Transformer-based cross reference network for video salient object detection. Pattern Recogn. Lett. **160**, 122–127 (2022). https://doi.org/10.1016/j.patrec.2022.06.006
24. Cong, R., Song, W., Lei, J., Yue, G., Zhao, Y., Kwong, S.: PSNet: parallel symmetric network for video salient object detection. IEEE Trans. Emerg. Top. Computat. Intell. **7**(2), 402–414 (2022). https://doi.org/10.1109/TETCI.2022.3220250

BCALI: A Blockchain-Based Lightweight Cross-Domain Access Control Model for IoT

Haonan Yang, Yue Sun, Wenqian Zhang, and Guanglin Zhang

1 Introduction

With the rapid development of IoT technology, a large number of IoT devices have been integrated into the Internet, greatly enhancing convenience [1]. However, this has also led to increased complexity in operations and data exchange. IoT devices typically face limitations in computing power, storage capacity, and network bandwidth, posing significant challenges to data management and security [2]. These constraints not only affect communication efficiency between devices but also introduce security risks such as data breaches and unauthorized access [3]. Therefore, implementing flexible and efficient solutions is crucial in the design and deployment of IoT systems.

Blockchain technology has garnered increasing attention in the IoT sector due to its decentralized, immutable, and transparent characteristics, offering potential enhancements in data security and privacy protection [4, 5]. Although blockchain-based autonomous cross-domain access control schemes for IoT have been proposed in recent years [6, 7], the growing number of IoT devices poses challenges. Specifically, the speed of processing large volumes of transactions and limited storage capacity on the blockchain are inadequate, hindering its widespread application in the IoT domain [4].

Some solutions attempt to store data in the cloud [8], which, while increasing storage capacity, also raises computational costs and the complexity of access strategies. Another approach is to store data locally, using the IPFS [9], a decentralized storage technology that allows distributed data storage. However, there are still

H. Yang · Y. Sun · W. Zhang (✉) · G. Zhang (✉)
Donghua University, Shanghai, China
e-mail: yhn@mail.dhu.edu.cn; sun_yue23@mail.dhu.edu.cn; wqzhang@dhu.edu.cn; glzhang@dhu.edu.cn

X. Jia et al. (eds.), *2nd International Conference on Big Data, IoT, and Cloud Computing*, EAI/Springer Innovations in Communication and Computing,
https://doi.org/10.1007/978-3-031-93570-1_14

shortcomings in cross-domain access control models. Several studies have sought to address IoT security challenges by combining blockchain and access control. For instance, Merlec et al. [10] proposed a context-aware access control based on smart contracts, while Zhang et al. [11] developed a smart contract-driven model for smart city applications. Fedrecheski et al. [12] introduced a simplified ABAC model executable on constrained IoT devices. In addition, there are several solutions in cross-domain access control [7, 13, 14].

Although existing methods like Role-Based Access Control and Attribute-Based Access Control [15] perform well in traditional systems, they lack flexibility in distributed IoT environments. Next-Generation Access Control [16] combines the strengths of RBAC and ABAC, offering improved policy definition and combination capabilities. However, implementation on resource-constrained IoT devices still faces computational and storage challenges. Additionally, effectively integrating decentralized storage systems like IPFS with access control mechanisms in IoT remains unresolved. Thus, innovative and efficient solutions and strategies are needed to enable effective access control and data storage.

To address the aforementioned issues, this chapter proposes a lightweight cross-domain access control model named BCALI. This model redesigns the cross-domain access control scheme and is based on consortium blockchain and IPFS. It uses smart contracts to achieve automatic permission verification and policy adjustment while leveraging the immutability of blockchain and the storage optimization of IPFS, thereby enhancing the efficiency, flexibility, security, and privacy protection of multi-device cross-domain management. The main contributions of this chapter are as follows:

- We design a lightweight cross-domain access control model for IoT systems, BCALI, which combines consortium blockchain and IPFS, enabling efficient management and access control in distributed and heterogeneous IoT environments.
- To ensure the security and scalability of the system, we propose a fine-grained cross-domain access control scheme that supports data interactions within different IoT domains.
- Based on BCALI, we implement a cross-domain data interaction access control process between users and devices through an interaction protocol designed using smart contracts.

2 System Model and Analysis

2.1 Decentralized System Model

To establish a secure and reliable IoT cross-domain access control system, we implemented a decentralized architecture. As shown in Fig. 1 the BCALI model

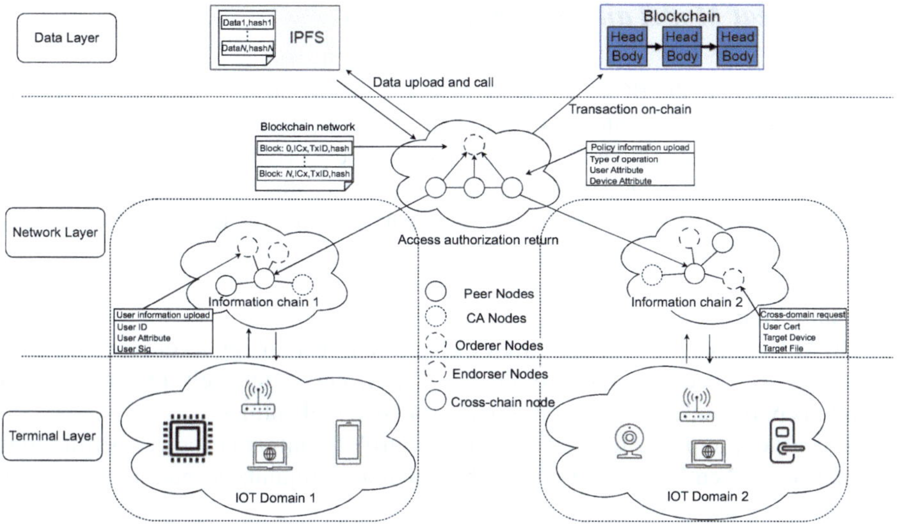

Fig. 1 Overall system architecture

consists of four components: blockchain, blockchain network, IPFS, and IoT domain.

Blockchain The blockchain is used to store transaction records and access logs, leveraging its immutability and traceability to ensure data integrity and transparency. Peer nodes handle transaction verification.

Blockchain Network Each IoT domain has a separate Hyperledger Fabric channel for node communication, which includes Orderer, Peer, CA, Cross-Chain, and Endorser nodes. The Orderer organizes transactions into blocks, the Peer nodes validate transactions and execute contracts, and the CA manages identities and certificates. Cross-chain nodes enable interoperability by managing inter-domain access, while Endorsers simulate and endorse transactions. Cross-domain transactions require consensus among cross-domain Endorsers.

IPFS *IPFS* provides distributed storage for large-scale data, integrated with the blockchain to uniquely hash IoT data. The blockchain records transaction data and hash values, while *IPFS* splits and encrypts the data across nodes to enhance scalability and security, minimizing reliance on centralized systems.

IoT Domain This includes gateways, IoT devices, and users. The IoT domain utilizes blockchain for secure data transactions. Through cross-domain access control, user requests are verified, and authorized access to *IPFS* data is granted. This ensures secure data flow and adaptive permission management in dynamic network environments.

2.2 Threat Model and Security Analysis

Threat Model In the BCALI system, potential threats include malicious third parties that may attempt to intercept communications or inject false data. Additionally, there are insider risks, such as employees with access privileges who might misuse them. Furthermore, attackers might carry out Sybil attacks by creating multiple fake identities to influence the consensus mechanism and compromise user privacy through transaction analysis.

Security Analysis
- BCALI utilizes the immutability of blockchain technology to ensure that once data is recorded, it cannot be altered. This feature effectively protects against interception and data injection threats from external entities.
- The system implements precise access controls, granting necessary permissions only to authenticated identities. This approach minimizes internal threats by preventing unauthorized access.
- By employing resource-based verification methods, BCALI ensures that only legitimate participants can partake in the consensus process. This reduces the risk posed by fraudulent identities and mitigates the impact of Sybil attacks.

3 Cross-Domain Access Control Scheme

The cross-domain access control solution consists of users (U) as access subjects and devices (O) as access objects, with user and device attribute sets (UA, OA) and access rights (ARS). Policy classes (PC) govern access, ensuring U and O remain independent with disjoint attribute sets.

To meet the demands of various IoT domains, the scheme includes access control domains (D). Each domain D_i defines specific users (D_iU), devices (D_iO), and policy classes (D_iPC). Cross-domain role mapping ($MA_i \subseteq D_jPC \times D_iPC$) enables role interoperability across domains, promoting seamless integration.

The BCALI scheme enhances NGAC by introducing differentiated access control domains and cross-domain role mappings (MA), organized into subsets $MA = MA_1 \cup MA_2 \cup \cdots \cup MA_n$. Mapping policy classes from domain D_j to D_i allows users meeting certain criteria to access the corresponding objects efficiently.

Access policies rely on three relationships: **Assignment**, **Association**, and **Prohibition**. **Assignment** creates a hierarchy linking user and device attributes, while **Association** grants access rights unless restricted. **Prohibition** clearly defines access boundaries, with policies visualized as directed acyclic graphs representing the interactive landscape.

Figure 2 displays the cross-domain access control in an IoT setting. Users are hollow ellipses linked to attributes—Admin, User, and Visitor—via solid lines, while devices are solid ellipses with functional or locational attributes. Dotted lines

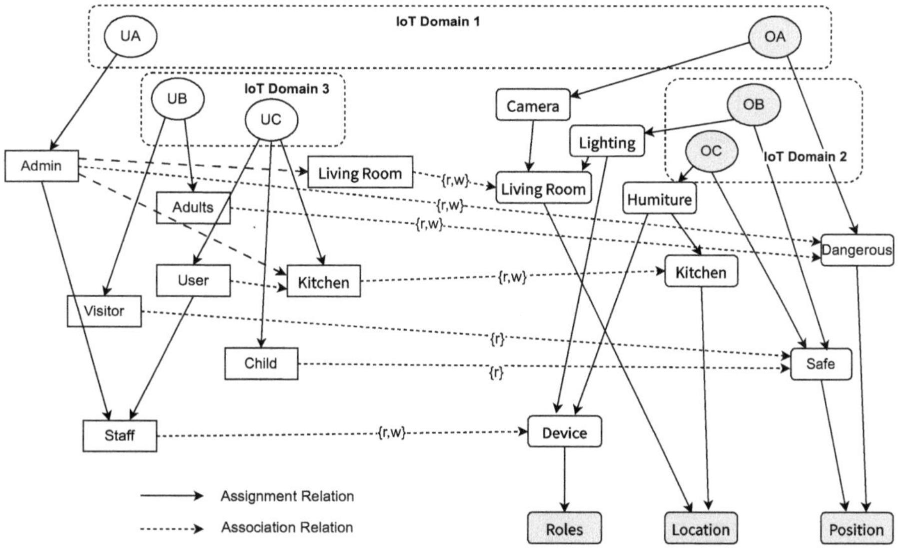

Fig. 2 Part of a cross-domain access control system

mark access rights, and policy classes (solid rectangles) include Role, Location, and Position policies. The "Role" policy grants role-based access, "Location" restricts users to areas, and "Position" gives read-only access for specific users or devices. The framework, emphasizing attribute-based permissions, ensures scalability and granularity. In multi-domain environments, each domain manages local policies only, with all changes recorded on the blockchain for auditable and secure operations.

4 Concrete Realization

The proposed solution involves several key entities: users, devices, information chain (IC), blockchain, and IPFS. These entities connect with each other to form a P2P blockchain network. For user U_A in domain D_1 to access or perform operations on data O_B in domain D_2, the system workflow is illustrated in Fig. 3 and is divided into several stages:

(1) User and Device Registration Phase: Steps 1–3
Step 1: $U_A \rightarrow IC_1 : \{ID_A, \text{sig}(sk_A, \text{times}), \text{times}\}$: User U_A sends an ID to the certificate authority IC_1 for registration, along with a signed timestamp for security.
Step 2: $IC_1 \rightarrow U_A : \{Cert_A\}$: IC_1 generates a certificate and signature, sending them to U_A. Based on U_A's information, IC_1 generates the device A's certificate and delivers it.

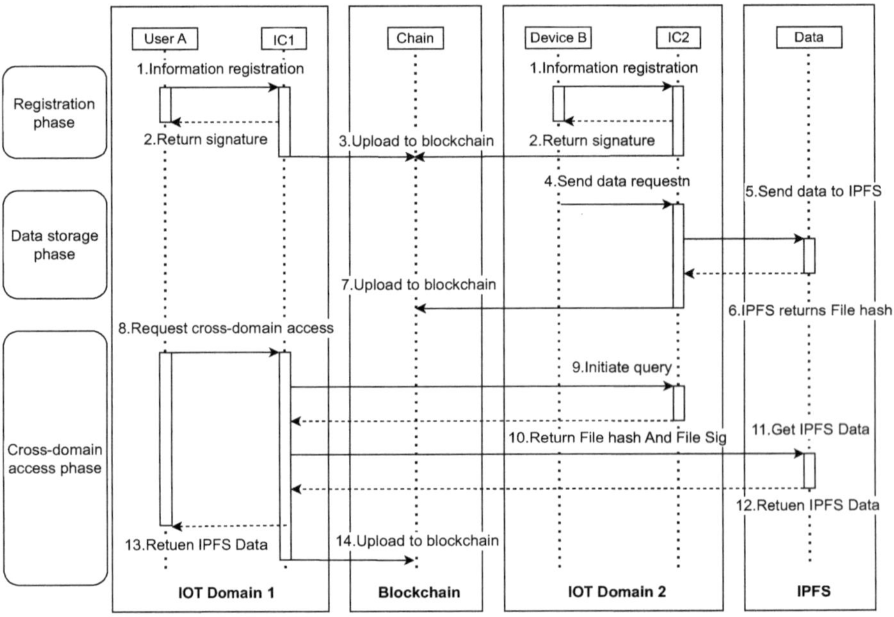

Fig. 3 Flow of cross-domain data access

Step 3: $U_A \rightarrow$ Blockchain : $\{Cert_A\}$: U_A uploads the certification to the blockchain to ensure transparency and immutability of identity information.

(2) Data Upload Phase: Devices O_B upload data to the blockchain, encompassing steps 4–7

Step 4: $O_B \rightarrow IC_2$: $\{Cert_B, Data_B, \text{sig}(sk_B, \text{times}), \text{times}\}$: Device O_B sends data to IC_2. IC_2 logs each data entry and updates the file signature $FileSig_x^y$, where x is the count and y the device ID. Upon reaching a preset value, IC_2 packages the data and streams it to IPFS.

Step 5: $IC_2 \rightarrow$ IPFS : $\{\text{Data}_x^B\}$: IC_2 sends a data block to IPFS, where the data block is the packaged FileSig_x^y.

Step 6: IPFS $\rightarrow IC_2$: $\{FileHash_x^B\}$: IPFS returns a file hash $FileHash_x^y$ to IC_2.

Step 7: $IC_2 \rightarrow$ Blockchain : $\{FileSig_x^B, FileHash_x^B\}$: Device O_B uploads the file signature and the returned file hash from IPFS to the blockchain. To prevent data bloat and potential security issues, IC_2 deletes the data blocks sent to IPFS.

(3) Cross-Domain Data Access Phase: Steps 8–14

Step 8: $U_A \rightarrow IC_1$: $\{\text{Cert}_A, D_2, O_B, x\}$: U_A sends a cross-domain request and authentication, specifying the IoT domain's device and file, where x indicates the operation.

Step 9: $IC_1 \rightarrow IC_2$: $\{U_A, O_B, x\}$: Upon receiving a data access request, if the requested IoT device is within domain D_1 (i.e., U_A and O_B belong to the same IoT domain), the process skips to step 11 where IC_1 requests the data from IPFS. Otherwise, IC_1 forwards U_A's information and the data request to IC_2. IC_2 then uses an access control model to verify if U_A has permission to access the file.

Step 10: $IC_2 \rightarrow IC_1$: {FileSig_x^B, FileHash_x^B}: If access is authorized, IC_2 returns FileSig_x^B and FileHash_x^B to IC_1.

Step 11: $IC_1 \rightarrow$ IPFS : {FileHash_x^B}: IC_1 sends FileHash_x^B to IPFS to retrieve the corresponding data.

Step 12: IPFS $\rightarrow IC_1$: {Data_x^B}: IPFS returns the data block to IC_1.

Step 13: $IC_1 \rightarrow U_A$: {Data_B}: IC_1 unpacks the data block using FileSig_x^B and sends it to U_A.

Step 14: $U_A \rightarrow$ Blockchain: U_A uploads this transaction to the blockchain.

5 Experiment and Analysis

5.1 Experimental Environment

In this section, we implemented a prototype to evaluate the feasibility and performance of the fine-grained data access control mechanism proposed in this chapter using the Hyperledger Fabric blockchain platform. Hyperledger Fabric is an open-source framework known for its high customizability and wide application in blockchain scenarios. The experimental setup included an Intel Core i7-12700 processor at 2.10 GHz, VMware Workstation Pro 16.2.4 for virtualization, and Ubuntu 22.04 LTS as the operating system, running on 4 GB RAM. We used Hyperledger Fabric version 2.2.0, IPFS version 0.25.0, Golang version 1.10, and Docker version 19.03.

Our blockchain network consisted of three organizations, each with its own peer nodes. This setup created independent information chains, enhancing system diversity and scalability. Various IoT devices were deployed, most notably ESP32-S3 microcontrollers that sent image data to the blockchain nodes over Wi-Fi. Additionally, STM32H743ZIT6-based devices were used to transmit file data using Ethernet and TCP protocols. The overall architecture of the experiment is shown in Fig. 4.

5.2 System Performance Analysis

In our experiment, we evaluated the impact of different devices on system performance by executing 50 transactions for each data packet size. During data transmission, IoT devices periodically send data. When a threshold is reached, the data is aggregated, encrypted using AES, uploaded to IPFS, and recorded on the

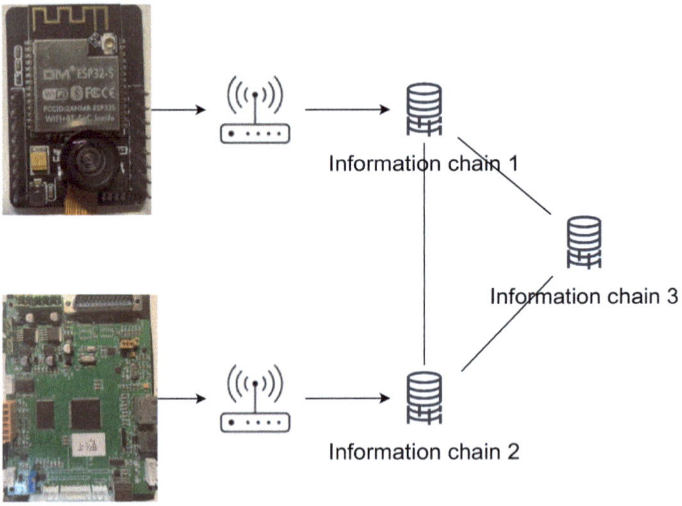

Fig. 4 Cross-domain access flow

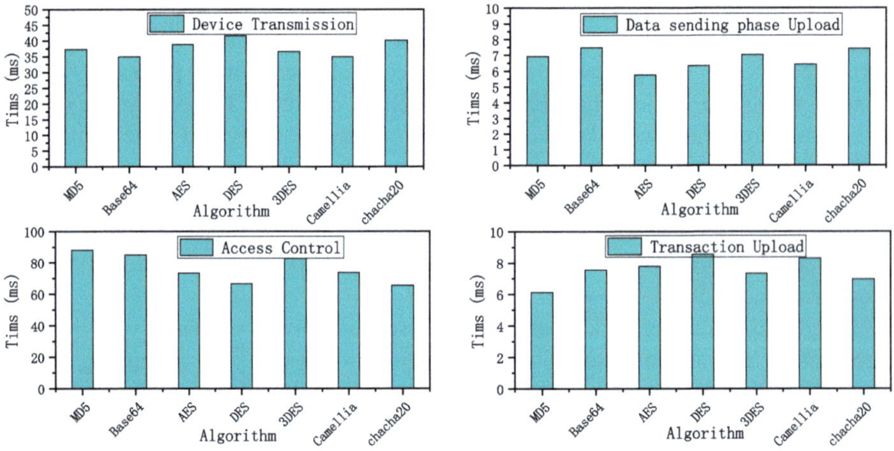

Fig. 5 Energy consumption of different encryption algorithms

blockchain. In cross-domain access, user requests trigger permission verification, retrieve files from IPFS, decrypt them, and log the transaction on the blockchain.

Figure 5 presents the evaluation of blockchain performance, where we tested various cryptographic schemes to assess the proposed system's performance on IoT devices. We measured the time taken for devices to send fixed-size data packets and upload transactions, averaging 38.36 ms and 6.63 ms, respectively. In the cross-domain access phase, the benchmark showed that permission checking and retrieval took 75.65 ms, and transaction logging took 7.25 ms. The results indicate that IoT devices can handle this workload.

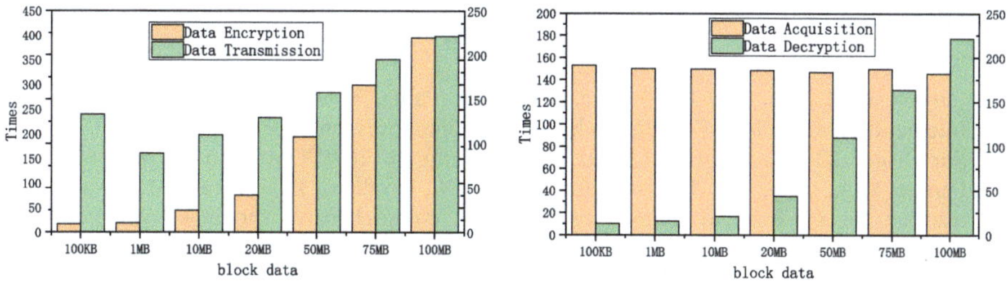

Fig. 6 Impact of data block size on system performance

Moreover, the data encryption and transmission time have a linear correlation with the data packet size, showing a time complexity of $O(n)$. Regardless of decryption time, cross-domain retrieval remains efficient and consistent. This analysis and comparative tests highlight our model's robustness and efficiency in supporting different IoT domains and devices, outperforming existing solutions in various metrics. Figure 6 further illustrates the results.

6 Conclusion

In this chapter, we proposed BCALI, a blockchain-based Lightweight Cross-Domain Access Control Model that was tailored for distributed and heterogeneous IoT environments. By integrating Hyperledger Fabric and IPFS, BCALI provided a fine-grained access control mechanism and a cross-domain data interaction protocol, ensuring secure and efficient management and data access for resource-constrained IoT devices. This significantly enhanced the model's scalability and flexibility. BCALI's innovative approach not only safeguarded data security and privacy but also promoted seamless interoperability among diverse IoT ecosystems. This made BCALI a promising solution for addressing contemporary challenges in the expanding IoT landscape.

We still face several challenges in the exploration of blockchain and cross-domain access control. One challenge is protecting device privacy; current software protection solutions perform poorly on IoT devices. Hardware-based protection schemes could be employed to enhance device privacy. Another challenge is the efficiency of blockchain. Currently, blockchain does not perform well in handling parallel tasks. High-performance computing could be utilized to optimize the system.

Acknowledgments This work was supported in part by the Shanghai Pujiang Programme (23PJD041), the Chenguang Program of Shanghai Education Development Foundation and Shanghai Municipal Education Commission (23CGA60), the AI-Enhanced Research Program of Shanghai Municipal Education Commission (SMEC-AI-DHUY-10), and in part by the Fundamental Research Funds for the Central Universities (2232025D-47).

References

1. Stoyanova, M., Nikoloudakis, Y., Panagiotakis, S., Pallis, E., Markakis, E.K.: A survey on the Internet of Things (IoT) Forensics: challenges, approaches, and open issues. IEEE Commun. Surv. Tutor. **22**(2), 1191–1221 (2020). https://doi.org/10.1109/COMST.2019.2962586
2. Ferrag, M.A., Friha, O., Kantarci, B., Tihanyi, N., Cordeiro, L., Debbah, M., et al.: Edge learning for 6G-Enabled Internet of Things: a comprehensive survey of vulnerabilities, datasets, and defenses. IEEE Commun. Surv. Tutor. **25**(4), 2654–2713 (2023). https://doi.org/10.1109/COMST.2023.3317242
3. Chen, C.-M., Chen, Z., Kumari, S., Obaidat, M.S., Rodrigues, J.J.P.C., Khan, M.K.: Blockchain-based mutual authentication protocol for IoT-enabled decentralized healthcare environment. IEEE Int. Things J. **11**(14), 25394–25412 (2024). https://doi.org/10.1109/JIOT.2024.3396488
4. Gharavi, H., Granjal, J., Monteiro, E.: Post-quantum blockchain security for the Internet of Things: survey and research directions. IEEE Commun. Surv. Tutor. **26**(3), 1748–1774 (2024). https://doi.org/10.1109/COMST.2024.3355222
5. Pérez-García, J.C., Braeken, A., Benslimane, A.: Blockchain-based group key management scheme for IoT with Anonymity of Group Members. IEEE Trans. Inf. Forens. Secur. **19**, 6709–6721 (2024). https://doi.org/10.1109/TIFS.2024.3414663
6. Hao, X., Ren, W., Fei, Y., Zhu, T., Choo, K.-K.R.: A blockchain-based cross-domain and autonomous access control scheme for Internet of Things. IEEE Trans. Serv. Comput. **16**(2), 773–786 (2023). https://doi.org/10.1109/TSC.2022.3179727
7. Tong, F., Chen, X., Wang, K., Zhang, Y.: CCAP: a complete cross-domain authentication based on blockchain for Internet of Things. IEEE Trans. Inf. Forens. Secur. **17**, 3789–3800 (2022). https://doi.org/10.1109/TIFS.2022.3214733
8. Yao, Y., Chang, J., Zhang, A.: Efficient data sharing scheme with fine-grained access control and integrity auditing in terminal–edge–cloud network. IEEE Int. Things J. **11**(16), 26944–26954 (2024). https://doi.org/10.1109/JIOT.2024.3393639
9. Benet, J.: IPFS - Content Addressed, Versioned, P2P File System. arXiv:1407.3561 (2014)
10. Merlec, M.M., In, H.P.: SC-CAAC: a smart-contract-based context-aware access control scheme for blockchain-enabled IoT systems. IEEE Int. Things J. **11**(11), 19866–19881 (2024). https://doi.org/10.1109/JIOT.2024.3371504
11. Zhang, Y., Yutaka, M., Sasabe, M., Kasahara, S.: Attribute-based access control for smart cities: a smart-contract-driven framework. IEEE Int. Things J. **8**(8), 6372–6384 (2021). https://doi.org/10.1109/JIOT.2020.3033434
12. Fedrecheski, G., De Biase, L.C.C., Calcina-Ccori, P.C., et al.: SmartABAC: enabling constrained IoT devices to make complex policy-based access control decisions. IEEE Int. Things J. **9**(7), 5040–5050 (2022). https://doi.org/10.1109/JIOT.2021.3110142
13. Zhang, S., Yan, Z., Liang, W., et al.: BCAE: a blockchain-based next-generation omain authentication scheme for edge computing. IEEE Int. Things J. **11**(13), 24035–24048 (2024). https://doi.org/10.1109/JIOT.2024.3387934
14. Feng, C., Liu, B., Guo, Z., et al.: Blockchain-based cross-domain authentication for intelligent 5G-enabled internet of drones. IEEE Int. Things J. **9**(8), 6224–6238 (2022). https://doi.org/10.1109/JIOT.2021.3113321
15. Ferraiolo, D., Chandramouli, R., Hu, V., et al.: A comparison of attribute based access control (ABAC) standards for data service applications. NIST Spec. Publ. **800**, 178, 1–68 (2016)
16. W. J. (ed.): Next Generation Access Control (NGAC). INCITS (2020).

Models of Adaptation of Cloud Technologies of Project Management in Organizational Industrial Enterprises

Murooj Zaiter, Mohammed Arnawtee, Majid Hamid Ali, and Ahmed Saadi Abdullah Al Basha

1 Introduction

The integration of cloud technologies into project management systems within industrial enterprises has become a focal point of research and innovation. As the industrial landscape evolves dynamically, there is a growing recognition of the need to adapt organizational processes to leverage the benefits of cloud technology [1]. This manuscript delves into the models of adaptation for cloud technologies in project management within industrial enterprises. By elucidating the opportunities and challenges inherent in this transformative process, our research aims to contribute to the enhancement of organizational efficiency, scalability, and security [2]. Through a comprehensive exploration of adaptation models, this paper seeks to provide valuable insights into the strategic implementation of cloud technologies in industrial settings [3].

M. Zaiter (✉) · M. Arnawtee
Department of Big Data Analysis and Methods of Analysis Institute of Radio Electronics and Information Technologies, Ural Federal University, Ekaterinburg, Russia

M. H. Ali · A. S. Abdullah Al Basha
College of Computer Science and Math, Tikrit University, Tikrit, Iraq

X. Jia et al. (eds.), *2nd International Conference on Big Data, IoT, and Cloud Computing*, EAI/Springer Innovations in Communication and Computing,
https://doi.org/10.1007/978-3-031-93570-1_15

2 Ease of Use

2.1 Understanding Cloud Technology

Cloud computing is the delivery of computing resources over the Internet. It allows users to access storage space, databases, software applications, and processing power from anywhere with an Internet connection and a device [4, 5]. Cloud computing offers cost-efficient and flexible business models, as companies can outsource these resources instead of purchasing expensive software and installing them on PCs. Cloud computing is not a completely new idea and has connections to technologies such as Grid Computing and distributed computing [6, 7]. It provides on-demand access to a shared pool of computing resources and enables ubiquitous access to data and services. The key characteristics of cloud computing include agility, cost efficiency, device and location independence, scalability, and security. Cloud computing has its own unique characteristics that distinguish it from other research areas, such as conceptual, advanced, monetary, and consumer experience [8, 9]. Cloud computing is a new technology that has the potential to transform data centers and provide a wide range of IT resource usage and application services under a converged network infrastructure [10, 11].

As shown in Fig. 1, cloud technology structure and components. The present study relies on the rich Google Cluster Usage Traces dataset that offers detailed operational metrics and task execution data from Google's cloud computing infrastructure [12, 13]. Traditional approaches in failure prediction often rely on rule-based systems or simple statistical models, which fall short of grasping complex patterns and interdependencies in cloud task behaviors [14].

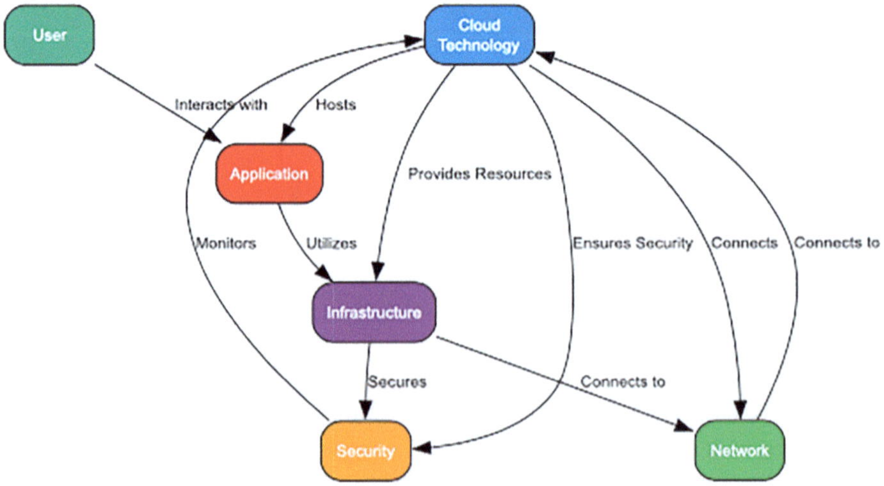

Fig. 1 Cloud technology structure and components

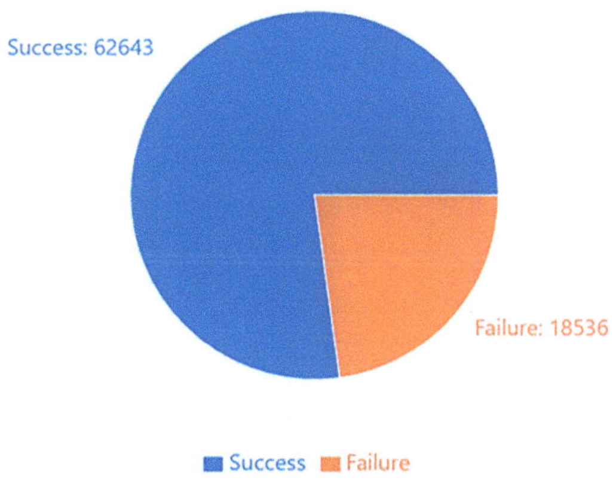

Fig. 2 Task outcome distribution

Our study mitigates this limitation by applying an advanced deep learning architecture with the integration of GRU layers, which uses state-of-the-art preprocessing to deal with the multi-dimensional nature of data generated by cloud computing. It takes in a variety of features such as resource requests, usage patterns, scheduling information, and system metrics to predict task failures [15, 16].

2.2 Google Cloud Cluster Usage Traces Dataset

The Google Cloud Cluster Usage Traces dataset presents a large, comprehensive sampling of operation metrics from the cloud computing infrastructure of Google [17].

As shown in Fig. 2 in total, 81,179 samples of cloud tasks, each characterized by 28 different features, are involved in detailed cloud task behavior and their failure patterns. The key metrics for this include CPU and memory usage, both raw values and distributions, and some advanced system metrics such as cycles per instruction and memory access patterns. Resource requests and actual usage patterns are stored as nested dictionary structures, which precludes straightforward analysis without careful preprocessing [18].

As shown in Fig. 3 task priority distribution. The dataset is imbalanced, with 77.17% of the tasks successfully completed and 22.83% failing, which makes it a very relevant dataset for studies on failure prediction. Temporal information is captured through timestamps and duration metrics, while system-level information includes machine IDs, instance indices, and cluster identifiers. In fact, some of the features, like CPU usage distributions and resource metrics, have a nested structure, introducing special challenges in preprocessing while carrying very valuable granular information for analysis. Its complexity and comprehensive nature make this dataset a perfect candidate for developing sophisticated failure prediction models in cloud computing environments.

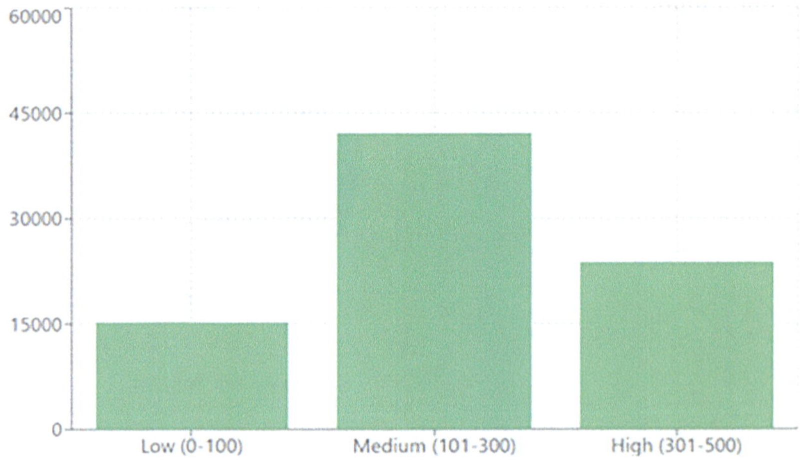

Fig. 3 Task priority distribution

3 Proposed GRU Model

Our proposed GRU model architecture implements a sophisticated deep learning approach for cloud task failure prediction [19]. The achieved model performance was quite astonishing with 100% on both training and validation after 30 epochs of training. The training history demonstrates fast convergence, with accuracy reaching 99.8% by epoch 5 and remaining stable afterward. The learning process was optimized using Adam optimizer with an initial learning rate of 0.001, supported by ReduceLROnPlateau for adaptive learning rate adjustment [20]. Proposed GRU Model Architecture Summary is shown in Table 1.

4 Results and Discussions

The confusion matrix also demonstrates flawless classification performance: there are 62,643 correctly classified non-failure cases and 18,536 correctly classified failure cases. The convergence was very fast, with the model achieving 99.8% accuracy in the first five epochs and remaining stable through the rest of the training.

As shown in Fig. 4, training and validation accuracy over epochs Accuracy plot presents the detail of the learning curve of our GRU model over the training period. What impresses most is the rapid climb in accuracy within the first five epochs, at 99.8% training accuracy and 99.9% validation accuracy.

Training and validation loss over epochs as shown in Fig. 5. Loss curves detail the optimization process and learning dynamics. The training loss starts off at 0.116, which reflects the high degree of uncertainty in the model's initial

Table 1 Proposed GRU Model Architecture Summary

Layer (Type)	Output Shape	Parameters	Connected To
Input	(None, 1, 28)	0	–
GRU layer 1	(None, 1, 128)	51,840	Input
BatchNorm 1	(None, 1, 128)	512	GRU layer 1
Dropout (rate: 0.3)	(None, 1, 128)	0	BatchNorm 1
GRU layer 2	(None, 64)	37,120	Dropout
BatchNorm 2	(None, 64)	256	GRU layer 2
Dropout (rate: 0.3)	(None, 64)	0	BatchNorm 2
Dense 1	(None, 32)	2080	Dropout
BatchNorm 3	(None, 32)	128	Dense 1
Dropout (rate: 0.2)	(None, 32)	0	BatchNorm 3
Dense 2	(None, 16)	528	Dropout
BatchNorm 4	(None, 16)	64	Dense 2
Dropout (rate: 0.1)	(None, 16)	0	BatchNorm 4
Dense 3 (output)	(None, 2)	34	Dropout
Layer (type)	Output shape	Parameters	Connected to

Fig. 4 Training and validation accuracy over epochs

predictions, whereas the validation loss is much lower at 0.001, indicating strong initial generalization. The dramatic drop in training loss within the first five epochs to 0.008 is indicative of the efficiency of our optimizer and learning rate settings. This is a steep decline without causing instability or overfitting, which is supported by the constant low validation loss.

Confusion matrix of proposed GRU model classifier. Confusion matrix visualization provides a very clear breakdown of the model performance concerning all the classes of prediction as shown in Fig. 6. What stands out is the class separation with

Fig. 5 Training and validation loss over epochs

Fig. 6 Confusion matrix of proposed GRU model classifier

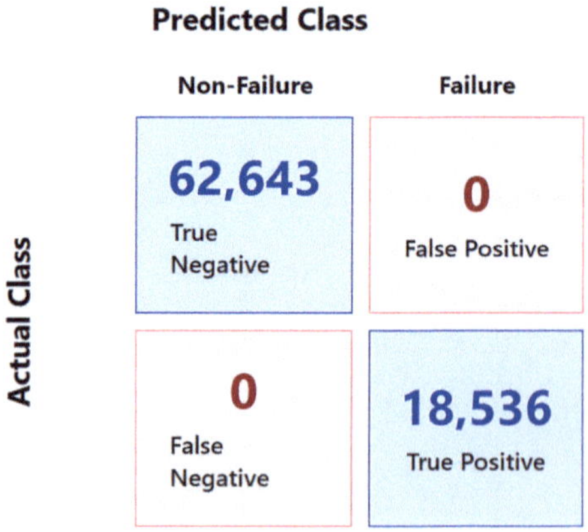

62,643 true negatives and 18,536 true positives with no instances of false positives and false negatives.

Performance metrics visualization represents the overall goodness of our model concerning several criteria. Perfect scores of 100% in accuracy, precision, recall, and F1-score demonstrate the high performance of this model. The accuracy metric also shows the correctness of the overall model predictions for both classes, with an accuracy of 100%. The precision score of 100% indicates zero false positives, crucial for maintaining operational efficiency by avoiding unnecessary interventions in cloud systems. The recall score of 100% demonstrates the model's capability to identify all actual failure cases, critical for maintaining system reliability and preventing service disruptions. The F1-score, being the harmonic mean of precision and recall, reaches 100%, validating the model's balanced performance across both positive and negative classes. The uniformity of these metrics suggests that the model maintains its performance across different operational scenarios and

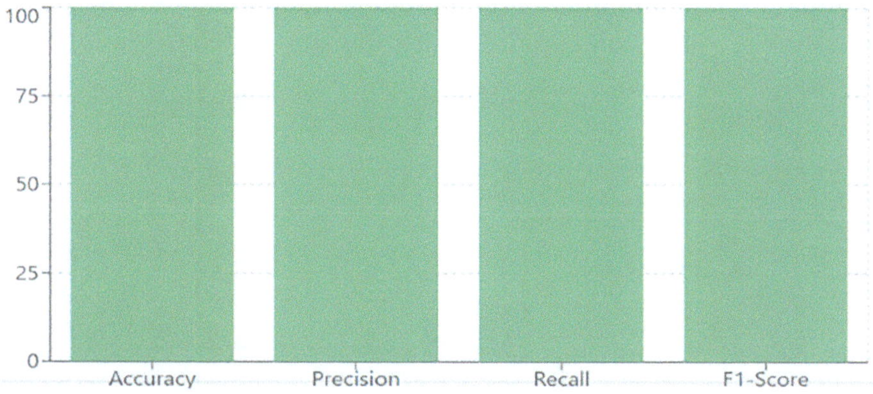

Fig. 7 Performance metrics visualization

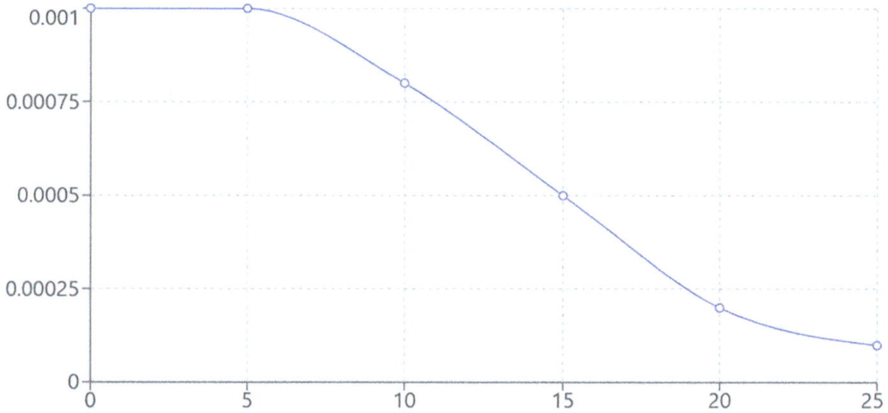

Fig. 8 Learning rate adaptation

failure patterns. This comprehensive excellence across all metrics validates our architectural decisions, training strategy, and the effectiveness of the GRU-based approach in handling the complexities of cloud computing workload data as shown in Fig. 7.

Learning rate adaptation curve provides a detailed insight into dynamic behavior within our optimizer through the entire training process, starting from an initial value of 0.001 to showcasing the power of our implemented ReduceLROnPlateau callback. The gradual step-wise reduction to 0.0001 over 25 epochs follows a carefully calibrated pattern: maintaining the initial rate during rapid early learning (epochs 0–5), followed by measured decreases at key convergence points (0.0008 at epoch 10, 0.0005 at epoch 15, and 0.0002 at epoch 20) as shown in Fig. 8.

Resource utilization visualization gives in-depth insight into the computational demands of our model during both training and inference phases. The consumption

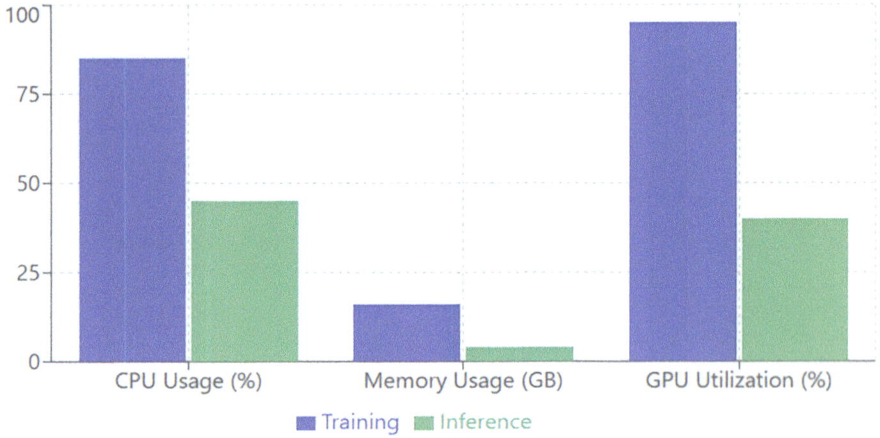

Fig. 9 Resource utilization comparison

of resources while training was intensive on all metrics: 85% CPU usage, reflecting the heavy nature of GRU computations; 16GB memory consumption, showing efficient batch processing and computation of gradients; and 95% GPU utilization, reflecting optimal hardware acceleration. In contrast, the inference phase exhibited significantly lower resource requirements of 45% CPU, 4GB memory footprint, and 40% GPU as shown in Fig. 9.

5 Conclusion

This research presents a huge milestone in cloud computing for task failure prediction by implementing a GRU-based deep learning model. Our architecture showed outstanding performance, classifying 100% of task failures on Google Cloud clusters with perfect precision and recall scores in all classes.

Its robustness is further validated by the model's ability to handle imbalanced data without explicit balancing techniques, maintaining stable performance across 77.17% success cases and 22.83% failure cases. The fast convergence to 99.8% within five epochs, combined with the sustaining of this performance throughout training, testifies to the efficiency of our learning strategy.

References

1. Anderson, C.: Cloud Computing: Concepts, Technology & Architecture. Morgan Kaufmann (2018)
2. Armbrust, M., et al.: A view of Cloud computing. Commun. ACM. **53**(4), 50–58 (2010)

3. Bernstein, D., et al.: Containers and Cloud: from LXC to Docker to Kubernetes. IEEE Internet Comput. **18**(2), 65–72 (2014)
4. Chou, D.: Cloud Computing: an Overview. CRC Press (2017)
5. Dikaiakos, M., et al.: Cloud computing: distributed internet computing for IT and scientific research. IEEE Internet Comput. **13**(5), 10–13 (2009)
6. Forbes, N., Wileman, A.: The Cloud Security Ecosystem: Technical, Legal, Business and Management Issues. Springer (2018)
7. Gartner, Magic Quadrant for Cloud Management Platforms. (2021). https://www.gartner.com/en/newsroom/press-releases/2021-07-28-gartner-reveals-2021-magic-quadrant-for-cloud-management-platforms
8. Hadoop.apache.org, Welcome to Apache Hadoop. (2021). https://hadoop.apache.org/
9. HashiCorp, Terraform—Cloud Infrastructure Automation. (2021). https://www.terraform.io/
10. IBM Cloud, IBM Cloud Computing. (2021). https://www.ibm.com/cloud
11. IEEE Cloud Computing. IEEE Cloud Computing. (2021). https://www.computer.org/cloud-computing
12. ISO/IEC JTC 1/SC 38: ISO/IEC 17788:2014, Cloud Computing—Overview and Vocabulary. ISO (2018)
13. Mell, P., Grance, T.: The NIST Definition of Cloud Computing. National Institute of Standards and Technology (2011)
14. Microsoft Azure, Microsoft Azure—Cloud Computing Services. (2021). https://azure.microsoft.com/
15. OpenStack, OpenStack—Open Source Cloud Computing Software. (2021). https://www.openstack.org/
16. Oracle Cloud Infrastructure, Oracle Cloud Infrastructure. (2021). https://www.oracle.com/cloud/
17. Salesforce, Salesforce—CRM and Cloud Computing To Grow Your Business. (2021). https://www.salesforce.com/
18. SAS, SAS Cloud Computing. (2021). https://www.sas.com/en_us/software/cloud.html
19. Shah, R., Hossain, M.: Handbook of Research on Cloud Computing and Big Data Applications in IoT. IGI Global (2017)
20. The Linux Foundation, Cloud Native Computing Foundation. (2021). https://www.cncf.io/

Semantic Communication-Based Task Offloading and Resource Allocation in Internet of Vehicles

Tao Hu and Xinjie Yang

1 Introduction

With the rapid development of intelligent transportation systems, the Internet of Vehicles (IoV) has quickly become one of the most important applications of mobile Internet of Things (IoT) technology [1]. In IoV systems, vehicles are typically equipped with sensing devices, such as cameras, to collect environmental information to support critical tasks such as autonomous driving, vehicle perception, and object detection. These tasks are often computationally intensive and latency-sensitive, demanding efficient utilization of computational resources and real-time responsiveness. Traditional resource allocation and task scheduling methods often suffer from inefficiencies in resource utilization and scalability. For example, they may allocate excessive resources to low-priority tasks while starving high-priority ones, resulting in a waste of computational power. Due to the limited computation resources of vehicles, many tasks cannot be effectively processed locally. The traditional solution is to offload tasks to centralized cloud servers. However, since cloud servers are normally distant from vehicles, this approach incurs significant transmission delays for task execution, leading to deteriorated Quality of Service (QoS). A commonly adopted alternative is to deploy edge computing servers with abundant computational resources in the Roadside Unit (RSU) and to offload vehicle tasks to the RSU for collaborative processing [2]. Before offloading video tasks,

T. Hu
Faculty of Electrical Engineering and Computer Sciences, Ningbo University, Ningbo, China

X. Yang (✉)
Faculty of Electrical Engineering and Computer Sciences, Ningbo University, Ningbo, China

Ningbo Arrowdeep Internet Technology Co. Ltd, Ningbo University, Ningbo, China
e-mail: yangxinjie@nbu.edu.cn

X. Jia et al. (eds.), *2nd International Conference on Big Data, IoT, and Cloud Computing*, EAI/Springer Innovations in Communication and Computing,
https://doi.org/10.1007/978-3-031-93570-1_16

vehicles typically perform simple preprocessing, while edge computing servers analyze and compute the offloaded video data, completing the computational and decision-making for various intelligent tasks [3].

In collaborative computing within the IoV, since vehicles generate a large number of computation offloading tasks, edge servers, due to their limited computation resources, inevitably need to queue tasks, regardless of the computation strategy employed. Tasks must first be stored in task queues, with their processing order determined by specific algorithms. Most existing studies determine tasks' priority based primarily on their arrival times at the service nodes. For example, Liu et al. investigate cloud-edge collaborative computing offloading scenarios and propose suitable queuing models for mobile devices, fog nodes, and cloud computing centers. Tasks are processed according to their arrival times at the service nodes, and the authors optimize users' offloading probabilities and transmission power using the interior point method to achieve better resource allocation [4]. Moreover, the authors of [5] consider the impact of both task data size and computational complexity on offloading priority. Using a hierarchical analysis model, tasks are prioritized to minimize processing delay. However, these strategies do not consider the importance and urgency of tasks, which means some important tasks may not be processed in a timely manner; hence, the Quality of Experience (QoE) is not delivered.

Although the aforementioned studies propose to offload tasks only to edge servers to compute, overlooking the computation resources of other network nodes in IoV networks. This may lead to server overload, hence negatively affecting system performance. To address this issue, some studies have explored utilizing idle vehicles to assist in processing computational tasks. For example, in [6], a joint offloading scheme is proposed that divides a task into three parts: one processed by the local vehicle, another offloaded to edge servers via V2I (Vehicle-to-Infrastructure) links, and the remaining part offloaded to idle vehicles via V2V (Vehicle-to-Vehicle) links. Similarly, Zhang et al. [7] introduce a binary task offloading strategy, leveraging idle vehicles and edge servers for collaborative task processing, and the total delay of tasks is optimized based on their priority. It is evident that incorporating the computation resources of idle vehicles not only alleviates the workload of edge servers but also significantly improves resource utilization and system flexibility.

In recent years, Semantic Communications (SC) has attracted increasing attention in academia, thanks to its potential to significantly enhance spectrum efficiency. It is expected to become a key technology for future mobile communications [8]. Traditional communication technologies are based on Shannon's classic information theory, which uses source and channel coding to ensure quality symbol reception at the receiver. In contrast, SC aims to ensure that the receiver accurately recovers the semantic information, focusing on the content beyond the symbols, and introduces new encoding methods and goals. Existing studies have demonstrated that using semantic encoding and decoding in some applications can effectively enhance the transmission efficiency at the physical layer [9]. Unlike traditional communication methods that transmit raw data, our scheme leverages semantic communication to extract and transmit only meaningful information. This not only reduces the bandwidth requirements but also enhances overall system efficiency by aligning the transmitted data with the computational needs of the task. In IoV systems, when video tasks are offloaded and processed through semantic encoding, network efficiency and user experience are no longer the sole performance metrics. For instance, improving the semantic accuracy becomes especially important. Therefore, it is necessary to develop a new resource allocation method that integrates video semantics with traditional QoS [10] and QoE [11], in order to more effectively meet the demands of intelligent tasks in IoV.

- Idle vehicles within the RSU's coverage area are leveraged as additional computing resources. Task offloading decisions are prioritized based on task importance to ensure that high-priority tasks are processed preferentially, enabling the joint optimization of objectives such as relative delay, social contribution rate, and semantic accuracy.
- Semantic communication is introduced into the task offloading and resource allocation process, emphasizing its benefits in reducing redundant data transmission and enhancing communication efficiency.
- An enhanced PSO algorithm tailored for IoV systems is proposed to optimize communication and computation resources. Compared to traditional optimization methods such as the genetic algorithm (GA) and ant colony optimization (ACO), the proposed algorithm significantly improves the system utility.

2 System Model

In this chapter, a mobile edge computing IoV system is considered, consisting of a single RSU and multiple vehicles, as shown in Fig. 1. An edge server is collocated with the RSU that has a coverage radius of R. Vehicles within the coverage area are equipped with on-board units (OBUs) that can provide computing services. It is assumed that the optimization interval is short enough so that during which vehicle position changes can be ignored. At the beginning of the interval, it is assumed that N vehicles initiate tasks, defined as User Vehicles (UVs), and there are M vehicles available to provide computation resources, defined as Idle Vehicles (IVs).

Vehicles perceive and acquire the surrounding environmental information through cameras and generate Semantic Information Tasks (SITs) after semantic preprocessing by OBU. The task generated by the u-th UV is denoted as Task u and represented as $h_u = \{d_u, c_u, t^{tole}, a_u, b_u, 1 \leq u \leq N\}$, where d_u is the task data size, c_u is the computing resources required to complete the task, t^{tole} is the delay constraint for the task, a_u represents the type of the task, and b_u is the type of the vehicle. Since UVs need to perform preprocessing such as video compression and semantic encoding on tasks, they generally do not have sufficient local computing resources to handle the generated tasks. Therefore, it is assumed that all UVs offload SITs to the RSU via the V2I links. Subsequently, the RSU then determines

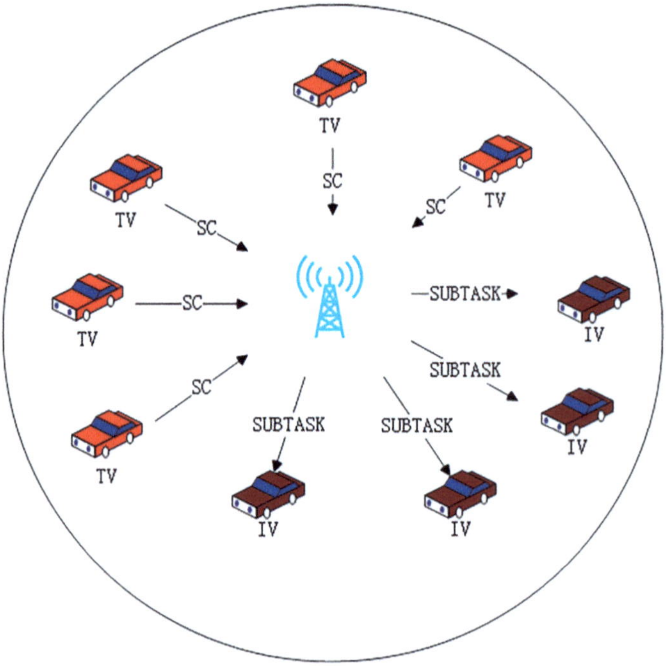

Fig. 1 System model

the priorities of the tasks based on a priority assessment model and processes one task at a time in order. Before processing tasks, the RSU first broadcasts a request, and IVs then respond by sending their vehicle information back to the RSU. The vehicle information of i-th IV is denoted as $P_i = \{D_i, f_i, 1 \leq i \leq M\}$, where D_i is the vehicle's distance from RSU and f_i the amount of computational resource the vehicle can provide for task offloading. Then the RSU decodes the task semantics and divides the task into M+1 subtasks according to the optimization algorithm. Among which, M subtasks are assigned to IVs via V2I links, while the RSU processes the remaining subtask. After the IVs complete their assigned subtasks, the results are sent back to the RSU. The RSU then sends the merged result back to the corresponding UV. For the sake of simplicity, assume that all tasks can be divided into any size, and each subtask has equal complexity and is independent with each other.

2.1 Priority Assessment Model

To enhance the user experience, an enhanced task priority assessment model based on the one in [12] is proposed. It additionally considers the weights of task importance and urgency. The RSU processes all tasks in sequence based on their priorities. The priority of the task h_u is defined as

$$p_u = \frac{(w_1 z_u + w_2 x_u - 1)(w_1 z_u + w_2 x_u - 2)}{2} + w_2 x_u \tag{1}$$

where z_u and x_u represent the importance and urgency of the task, respectively. w_1 and w_2 denote their corresponding weights. Similar to [12], the urgency of a task is defined as

$$x_u = \frac{c_u}{t^{tole}} \tag{2}$$

The importance of a task is determined by the task type and the vehicle type, expressed as

$$z_u = \frac{(a_u + b_u - 1)(a_u + b_u - 2)}{2} + b_u \tag{3}$$

where a_u is the task type and ranges in [1, 2, 3], representing ordinary, important, and emergency tasks, respectively. b_u is the vehicle type and ranges in [1, 2, 3], corresponding to ordinary, public service, and emergency vehicles, respectively.

2.2 Communication Model

In the studied IoV system, both the task offloading from UVs to RSU and the distribution of subtasks from RSU to IVs utilize V2I links. It assumes that the IoV system complies with the international standard LTE-V2X (Long-Term Evolution—Vehicle to Everything) system, and the Orthogonal Frequency-Division Multiple Access (OFDMA) technology is applied. The system allocates frequency-divided channels to each node for data transmission, consisting of multiple subcarriers. Due to the orthogonality between subcarriers, the interference between channels is minimal and can be neglected. When vehicle u offloads the task, the uplink rate and the received Signal-to-Noise Ratio (SNR) at the RSU can be defined, respectively, as

$$r_{u,R} = B_{u,R} \log_2(1 + \mathrm{SNR}_{u,R}) \tag{4}$$

$$\mathrm{SNR}_{u,R} = \frac{S_{UV} \left| h_{u,R} \right|^2 D_{u,R}{}^{-\alpha}}{N_0} \tag{5}$$

where the transmit power of all UVs is assumed the same, denoted as S_{UV}, and $D_{u,R}$ represents the distance between vehicle u and the RSU. $B_{u,R}$ is the allocated uplink channel bandwidth. $h_{u,R}$ is the coefficient of the Rayleigh fading channel, which follows the complex Gaussian distribution $CN(0, 1)$. a is the path loss exponent, and N_0 is the power of the background noise. When distributing subtasks, the downlink rate and the received SNR at the IV have similar definitions. To ensure transmission efficiency for different tasks, the RSU employs optimization algorithm to dynamically allocate the uplink bandwidth. In the downlink, considering that the amount of subtask data is relatively small, a fixed bandwidth allocation method is adopted to simplify the system design and improve the stability. The downlink channel bandwidth is defined as

$$B_{R,i} = \frac{B_R}{M} \tag{6}$$

where B_R represents the total bandwidth reserved by the RSU for subtask distribution.

2.3 Computation Model

Based on the system model, the total delay of h_u consists of three parts: (1) the transmission delay for the vehicle to offload the task to the RSU, (2) the computation delay for completing subtasks, including the transmission delay for subtasks distribution, (3) the queuing delay of the task at the RSU. Considering that the amount of data after subtask processing is very small, the transmission delay for returning the computation results can be ignored. Therefore, the transmission delay of h_u can be expressed as

$$t_u{}^{trans} = \frac{d_u}{r_{u,R}} \tag{7}$$

The computation delay of the task can be expressed as

$$\begin{aligned} t_{u,R}{}^{comp} &= \frac{\theta_{u,R} c_u}{f_R} \\ t_{u,i}{}^{comp} &= \frac{\theta_{u,i} c_u}{f_i} + \frac{\theta_{u,i} d_u}{r_{R,i}} \\ t_u{}^{comp} &= \max(t_{u,R}{}^{comp}, t_{u,1}{}^{comp}, \ldots, t_{u,M}{}^{comp}) \end{aligned} \tag{8}$$

where $\theta_{u,R}$ and $\theta_{u,i}$ are the subtask ratios of the RSU and IV determined by the optimization algorithm, respectively.

Let the queuing number of h_u in the RSU task queue be denoted as q_u; then the queuing delay of h_u is

$$t_u{}^{wait} = \begin{cases} 0 & , q_u = 1 \\ \sum_{j=1}^{q_u-1} t_j{}^{comp} & , q_u > 1 \end{cases} \tag{9}$$

The total delay is then

$$t_u^{\ total} = \max t_u^{\ trans} + t_u^{\ comp} + t_u^{\ wait} \tag{10}$$

The relative delay of h_u is defined as

$$T_u = \begin{cases} \frac{t_u^{tole} - t_u^{total}}{t_u^{tole}} & , \text{if } t_u^{tole} \geq t_u^{total} \\ 0 & , \text{otherwise} \end{cases} \tag{11}$$

In reality, the higher priority a task has, the greater it contributes to the society upon the completion. In this chapter, all tasks are divided into three priority groups, namely high, medium, and low, denoted as HPG, MPG, and LPG, respectively. Tasks in these priority groups result in different social contributions SC_1, SC_2, and SC_3 when completed within the delay constraint, respectively. The overall social contribution rate is defined as

$$SCR = \frac{SC_1 \times tn_{HPG}^{\text{succ}} + SC_2 \times tn_{MPG}^{\text{succ}} + SC_3 \times tn_{LPG}^{\text{succ}}}{SC_1 \times tn_{HPG}^{total} + SC_2 \times tn_{MPG}^{total} + SC_3 \times tn_{LPG}^{total}} \tag{12}$$

When the system experiences poor communication conditions or insufficient computation resources, tasks with higher importance should be completed first. To address this, to guide the system to complete important tasks first, an importance reward function is defined as

$$Y_u = z_u^{\ b} \tag{13}$$

where b is a coefficient greater than 1.

When vehicle u performs compression on the original video, it assumes that adopts the HEVC (High-Efficiency Video Coding) encoding scheme is employed, which is the H.265 video compression standard [8]. During the compression process, different video Quantitative Parameters (QPs) correspond to different compression steps. A smaller QP value results in a smaller compression step, which results in a transmission rate closer to the SIT transmission rate and requires more communication resources. Therefore, different transmission rates correspond to different QP and compression steps. According to Ref. [9], the QP can be expressed as

$$Q_{u,R} = \alpha_1 \exp(\alpha_2 r_{u,R}) \tag{14}$$

where α_1 and α_2 are the QP model parameters, and their optimal values vary depending on how the videos are captured. By statistically analyzing experimental data corresponding to different QP and their semantic accuracy, the relationship between semantic accuracy and QP [10] can be expressed as

$$P_u = \beta_1 (Q_{u,R})^{\beta_2} + \beta_3 \tag{15}$$

where β_1, β_2, and β_3 are model parameters, and their optimal values vary for different objects the video tries to detect, e.g., pedestrian, vehicles, and traffic lights.

3 Problem Formation and Description

In a nutshell, considering the relative delay, semantic accuracy, task importance reward, and social contribution rate of all tasks, a new system utility function is proposed and defined as

$$\text{Utility} = \Big(\frac{1}{N}\sum_{\text{n}\in N} T_{\text{u}} \times P_{\text{u}} \times Y_{\text{u}}\Big) \times SCR \tag{16}$$

The optimization objective of this study is to obtain the optimal uplink bandwidth allocation and task offloading ratio for all the tasks to maximize the system utility function. Therefore, the optimization problem can be described as

$$\begin{aligned}
\text{P1:}\ & \max_{B_{u,R},\theta_{u,i}} \ \text{Utility} \\
\text{s.t.}\ & C1: B_{u,R} \geq B_{\min} \\
& C2: \sum_{u=1}^{N} B_{u,R} \leq B_{\max} \\
& C3: \theta_{u,i} > 0 \\
& C4: \theta_{u,R} > 0 \\
& C5: \theta_{u,R} + \sum_{i\in M} \theta_{u,i} = 1
\end{aligned} \tag{17}$$

where constraint C1 ensures that the bandwidth allocated to each UV is not lower than the minimum bandwidth threshold $B_{\min}$. C2 ensures that the sum of the bandwidths allocated to all UVs does not exceed the maximum bandwidth of the system $B_{\max}$. C3 and C4 ensure that all IVs and RSU participate in the computation of subtasks. C5 ensures that the task can be completely offloaded and processed.

4 The Offloading Scheme Based on PSO Algorithm

Problem P1 is a nonlinear programming problem that cannot be solved by traditional convex optimization algorithms. For such problems, intelligent swarm algorithms can obtain relatively good solutions by conducting heuristic searches of the solution space. An enhanced PSO task offloading strategy based on a priority model that comprehensively considers importance and urgency is designed to solve P1 and named ePSOO-AIU (Enhanced Particle Swarm Optimization Offloading According to task Importance and Urgency).

4.1 The Settings of Particles

The optimization objective is to find the optimal bandwidth allocation and subtask ratios so as to maximize the utility function. Let the number of particles be L. In the initial state, the position of each particle is randomly generated according to constraints C1–C5. Then, a matrix with N rows and (1+M+1) columns is formed, where the first column of each row represents the bandwidth allocated to all UVs, and the subsequent M+1 columns represent the subtask ratios of IVs and RSU.

4.2 Fitness Function

The system utility function of Problem P1 is taken as the fitness function of the algorithm, and Y represents the value of the PSO algorithm's fitness.

$$Y = \text{Utility} = \left(\frac{1}{N} \sum_{\text{n} \in N} T_{\text{u}} \times P_{\text{u}} \times Y_{\text{u}} \right) \times SCR \tag{18}$$

4.3 Details of the Algorithm

In the PSO algorithm, the positions of the particles represent a feasible solution. During the iterations, all particles search for the optimal path by calculating their next positions until they converge to the best positions. Without loss of generality, for any particle, its position update depends on its previous position and its current velocity. The update of velocity is affected by three aspects, namely the inertial velocity, self-cognitive experience, and social experience. They are described by the inertial factor w and the learning factors c_1 and c_2, respectively. In this chapter, an enhanced PSO is proposed, where the performance of the algorithm is optimized by dynamically adjusting the values of these factors during the iteration process. The update formula for the velocity V^{t+1} of the particle at the (t+1)-th iteration is

$$\begin{aligned} V^{t+1} &= w^t \times V^t + c_1^t \times rand() \times (P_{best}^t - V^t) \\ &+ c_2^t \times rand() \times (G_{best}^t - V^t) \end{aligned} \tag{19}$$

where t represents the number of iteration, P_{best}^t is the current optimal solution of the particle, and G_{best}^t is the current optimal solution of all particles. To ensure that the algorithm has global convergence ability in the initial stage and local convergence ability in the later stage, the value of the inertia factor will gradually decrease during the iterations. The update follows

$$w^t = w_{\max} - \left(\frac{w_{\max} - w_{\min}}{T_{\max}}\right) \times t \tag{20}$$

where $T_{\max}$ is the maximum number of iterations, and $w_{\max}$ and $w_{\min}$ are the maximum and minimum values of the inertia factor, respectively. This update model ensures that in the initial stage of the algorithm, the value of w is relatively large, enabling the particles to conduct more extensive searches in the solution space.

The specific update formulas for c_1 and c_2 are as follows:

$$\begin{aligned} c_1^{t+1} &= c_1^t \cdot \left(1 + \eta \cdot \frac{|Y_{Pbest}^t - Y^t|}{1+t}\right) \\ c_2^{t+1} &= c_2^t \cdot \left(1 + \lambda \cdot \frac{Y_{Gbest}^t - Y^t}{Y_{Gbest}^t} \cdot \sqrt{1+t}\right) \end{aligned} \tag{21}$$

where Y^t, Y_{Pbest}^t, and Y_{Gbest}^t represent the fitness of the particle, the fitness of the particle's current optimal solution, and the fitness of the global optimal solution, at the t-th update, respectively. η and λ are adjustable coefficients. Using the updates, the search efficiency of the algorithm is improved, as the global and local search capabilities can be balanced, which helps to find the optimal solution more quickly.

Therefore, the position update for the l-th particle is

$$X_l^{t+1} = X_l^{t} + V_l^{t+1} \tag{22}$$

5 Simulation Results

The system-level simulation is based on an IoV system composed of a single RSU, multiple UVs, and multiple IVs as shown in Fig. 1. The simulation results are obtained by averaging 1000 snapshots. The key simulation parameters and settings are presented in Table 1.

To comprehensively evaluate the proposed AIU task priority model's effectiveness with the offloading strategy, performance comparisons are made with common priority models: ePSOO—ADC (prioritizes tasks with smaller delay constraints), ePSOO—ACS (prioritizes tasks needing fewer computational resources), and ePSOO—ATO (prioritizes tasks arriving first).

Additionally, to further assess the effectiveness and robustness of the proposed ePSOO—AIU algorithm, its performance is compared with common offloading algorithms in literature: GAO and ACO .

Table 1 Simulation parameters

Vehicular environment parameters	Value
Number of UVs within RSU range/vehicle	10
Number of IVs within RSU range/vehicle	5
Upload transmit power of UVs S_{UV}/w	0.2
Total bandwidth of downlink B_R/MHz	3
Delay constraint of UVs t^{tole}/s	[0.5, 3]
Data size of UVs d_u/KB	[100, 350]
Computation amount required by tasks c_u/GHz	[0.5, 1.5]
Computing capacity of RSU f_R/GHz/s	5
Computing capacity of IVs f_i/GHz	[1, 2]
Communication radius of RSU/m	500
Power of Gaussian white noise N_0/dBm	−80

Figure 2a shows the system utility of each task priority model versus distances between the UV and RSU. When the distance increases, the utility of all priority models decreases. This is because when the UV moves away from RSU, the SNR at RSU decreases, leading to a consequent reduction in the effective transmission rate and an increase in the task offloading time, which ultimately results in a lower system utility. In addition, ePSOO-AIU outperforms other three task priority models. This is because the ePSOO-AIU model takes into account the importance and urgency of tasks, so that the resource is preferentially allocated toward tasks with greater social contributions and importance rewards, effectively enhancing the system efficiency.

Figure 2b presents the comparison of system utility among ePSOO-AIU and two other common algorithms under different total uplink bandwidth. It can be observed that the utility of ePSOO-AIU is significantly higher than those of the GA and ACO algorithms. This is because ePSOO-AIU has optimized the inertia weight parameter and learning factors of the traditional PSO algorithm, thereby enhancing its optimization ability. The GA is able to perform operations such as selection, crossover, and mutation on chromosomes, leading to a performance second to that of ePSOO-AIU. The ACO algorithm generates solutions in a probabilistic manner by simulating the accumulation and evaporation of pheromones during the process of ants foraging. However, due to the slow update of pheromones and its proneness to getting trapped in local optimization, its performance is rather poor and significantly lower than the GA algorithm and ePSOO-AIU. In conclusion, ePSOO-AIU demonstrates superiority in system performance and stronger adaptability for task offloading in IoV systems.

Figure 2c shows the system performance of the three algorithms under different numbers of IVs. The results indicate that as the number of IVs increases, the utility of each algorithm increases. This is because a greater number of vehicles imply more nodes for subtask distribution and more computing resources, which can further enhance the system utility. Similar to the reasons explained in Fig. 2b, the system utility of ePSOO-AIU is significantly superior to other two algorithms.

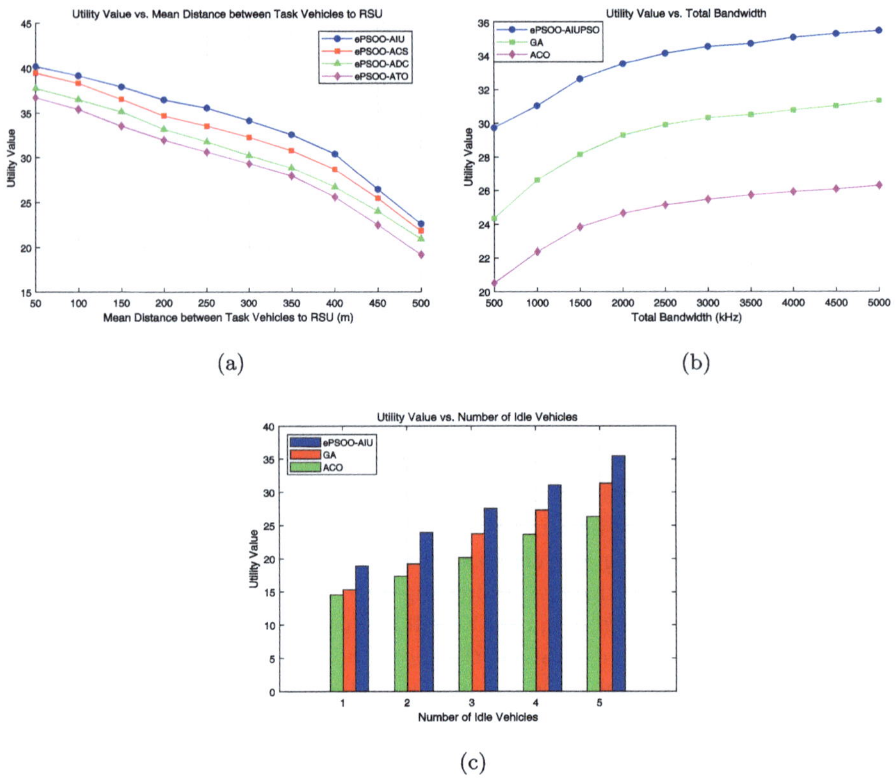

Fig. 2 Simulation results. (**a**) Utility of different vehicle distances of each priority model. (**b**) Comparison of utility values of each algorithm under different uplink bandwidths. (**c**) Comparison of utility values of each algorithm under different numbers of IVs

6 Conclusions

In this chapter, a novel task offloading strategy for mobile edge computing IoV systems is studied. A priority model based on the urgency and importance of tasks is proposed, and a task offloading strategy named ePSOO-AIU is constructed. By jointly optimizing the allocation of communication resources and computing resources of RSU and IVs, the utility of the system is maximized. The utility function comprehensively takes into account factors such as semantic accuracy, relative delay, importance reward, and social contribution rate. The PSO algorithm is used to solve this optimization problem, with improvements made to the traditional PSO's inertia weight and learning factors. Simulation results demonstrate that the proposed priority-based offloading algorithm significantly improves the overall system utility compared to traditional algorithms.

References

1. Zhuang, W., Ye, Q., Lyu, F., et al.: SDN/NFV-empowered future IoV with enhanced communication, computing, and caching. Proc. IEEE **108**(2), 274–291 (2019)
2. Fan, W., Han, J., Yao, L., et al.: Latency-energy optimization for joint WiFi and cellular offloading in mobile edge computing networks. Comput. Netw. **181**, 107570 (2020)
3. Wang, J., Feng, D., Zhang, S., et al.: Computation offloading for mobile edge computing enabled vehicular networks. IEEE Access **7**, 62624–62632 (2019)
4. Liu, Q., Mo, R.-C., Xu, X.-L., et al.: Multi-objective resource allocation in mobile edge computing using PAES for Internet of Things. Wirel. Netw. **26**(3), 1–13 (2020)
5. Zhao, H.-T., et al.: Research on content-aware classification offloading algorithm based on mobile edge calculation in the Internet of Vehicles. J. Electron. Inf. Technol. **42**(1), 20–27 (2022)
6. Fan, W., Liu, J., Hua, M., et al.: Joint task offloading and resource allocation for multi-access edge computing assisted by parked and moving vehicles. IEEE Trans. Vehic. Technol. **71**(5), 5314–5330 (2022)
7. Zhang, H., Wang, Z., Liu, K., et al.: V2X offloading and resource allocation in SDN-assisted MEC-based vehicular networks. China Commun. **17**(5), 266–283 (2020)
8. Jiujiu, C., Chunyan, F., Caili, G., et al.: Video semantics driven resource allocation algorithm in Internet of Vehicles. J. Commun. **42**(07), 1–11 (2021)
9. Chuanhong, L., Caili, G., Yang, Y., et al.: Semantic communication for intelligent tasks: theories, technologies and challenges. J. Commun. **43**(06), 41–57 (2022)
10. Li, G.Y., Boukhatem, L., Wu, J.S.: Adaptive quality-of-service-based routing for vehicular ad hoc networks with ant colony optimization. IEEE Trans. Vehic. Technol. **66**(4), 3249–3264 (2017)
11. Liang, H.B., Zhang, X.H., Hong, X.T., et al.: Reinforcement learning enabled dynamic resource allocation in the Internet of vehicles. IEEE Trans. Ind. Inf. **17**(7), 4957–4967 (2021)
12. Jing, T., Sheng, M., Wang, X., et al.: Enhancing soft Ac based reliable offloading for IoV with edge computing. IEEE Trans. Commun. **64**(10), 4268–4282 (2023)

Research on the Causal Relationship and Visualization of Drugs and Crime Driven By Big Data

Shaobing Wu and Changmei Wang

1 Introduction

There are currently many literature reviews on the causal relationship between drugs and crime [1]. Overall, these authors found that although a large amount of covariance can be recorded at most stages of drug addiction, there is some ambiguity in the "causal relationship direction" between drug use and criminal activity, and considerable ambiguity in specific antecedents and concurrent factors of influence. Based on this relationship, previous studies have shown that there are three perspectives: (1) the "reasons" for drug use or the triggering of crime (especially property crimes); (2) having a criminal tendency to promote and maintain drug use; (3) there is a false correlation between drug use and crime, as they depend on common triggering factors (i.e., socio-cultural or personal determinants). However, few suitable methods have been proposed or implemented to provide a basis for evaluating the relative advantages of these alternative solutions or specifying conditions that can be proven individually or in combination.

To the best of the author's knowledge, equation modeling methods have not yet been attempted in the fields of drug use and property crime. The complexity of existing research is almost limited to comparing the means, rates, and/or proportions of various criminal behaviors at different time points. Elliott et al. [2] used path analysis to investigate the impact of crime on criminal behavior and the etiology

S. Wu
Key Laboratory of Intelligent Drug Control, Yunnan Police college, Kunming, China

Institute of Information Security, Yunnan Police College, Kunming, China

C. Wang (✉)
Education Ministry Key Laboratory of Advanced Technology and Preparation for Renewable Energy Materials, Yunnan Normal University, Kunming, China

X. Jia et al. (eds.), *2nd International Conference on Big Data, IoT, and Cloud Computing*, EAI/Springer Innovations in Communication and Computing,
https://doi.org/10.1007/978-3-031-93570-1_17

of drug use. However, their model once again does not allow for determining the causal advantage between drug use and crime.

In structural equation modeling, latent variables are similar to factors in factor analysis because they are unobserved hypothetical structures represented by multiple measured or explicit variables. These potential structures have greater breadth and theoretical universality than the reference objects they observe, and therefore may be more useful in empirically testing general theoretical propositions. Especially, such testing allows for the unraveling of the causal relationship between two interdependent structures over time [3].

2 Related Work

Causal relationship is the general relationship between the result and the cause that produces the result. It is difficult to define, and we usually only rely on intuition to know the causal relationship. When it comes to learning causal relationships in data, we need to be aware of the difference between statistical associations and causal relationships.

The ability to learn causal relationships is considered an important component of human intelligence and can serve as the foundation of artificial intelligence. From a historical perspective, studying causal relationships has been studied in numerous high impact fields, including education [4], medicine [5], and economics [6]. Due to the limitation of data volume, solid prior causal knowledge is a necessary condition for learning causal relationships. Researchers study data collected through carefully designed experiments, where solid prior causal knowledge is crucial.

There are many other comprehensive investigations in the field of causal learning. Pearl [7] aims to convey the fundamental theory of causal relationships based on structural causal models. Gelman [8] proposed a high-level perspective on the existing formal frameworks and issues of causal learning.

3 Causal Relationship

3.1 Overview of Causal Relationships

From the moment we start observing and interacting with nature, we live in a constantly changing environment. We know that our actions will lead to consequences. Early personal experiences inferred these basic frameworks from the external world and quickly developed a simple and mechanical explanation of the world: causes produce results over time. This is our intuitive and practical model of reality.

Causal relationships, like other concepts of the world, are closely related to the existence of external observers and the ability of observers to interact with the world and change its processes.

In a causal environment, intervention or experimentation is defined as an external forcing variable that pushes it beyond its natural behavior, thereby cutting it out of naturally dependent variables. By observing the consequences of such behavior, we can understand the causal effects of the intervened variable. Intervention not only isolates a variable from naturally observable causes, but also isolates it from hidden confounding factors that can create causal relationships between two or more observed variables.

3.2 *Representation of Causal Relationships*

There are two common ways to represent causal relationships: causal diagrams and causal equations. In a causal diagram, vertices represent variables and edges represent causal relationships between variables. Use causal diagrams to encode causal relationships between variables. Both observable and unobservable variables can be represented as vertices in the graph, and causal relationships can be represented as edges. Causal relationship has directionality, that is, acting on one variable causes another variable to change; otherwise, it does not hold. A simple method to establish a causal relationship model is to use a directed acyclic graph (DAG). The causal equation is a causal relationship equation that is first determined by determining the parameters of causal variables and mining them based on experience, intuition, and relevant data.

3.3 *Discovering Causal Relationships*

Causal diagrams encode the causal relationships between variables in the model. By observing a group and attempting to build a causal graph from the observed features of the data. In a deterministic world, once we know the causal variable, we undoubtedly know the outcome.

Discovering causal relationships from empirical data is the foundation of scientific methods. In the past few decades, some focus has been placed on the development of algorithms for discovering causal relationships from observed probability data. The attempt to automatically discover causal relationships from purely observed data is aimed at avoiding the need for experimentation in the field, which can be a very expensive process. The systematic analysis of observed data under various assumptions provides appropriate hints, indicating that there may be causal relationships between the analyzed variables in the model.

The discovery of causal relationships from observational data has its limitations. In most cases, we can only obtain at most one set of causal models that match the observed data. Another approach is to not only identify causal relationships from observational data, but also to conduct experiments (also known as "interventions") on the variable system from which we attempt to discover causal relationships, and collect the joint probability distributions of the system under various interventions to achieve this. Establishing a complete directed acyclic graph may require many intervention measures.

4 Causal Relationship of Drug Crimes Based on Python

4.1 Algorithm

Bnlearn is a library in Python used for Bayesian network learning. It is based on pgmpy and supports structure learning, parameter learning, inference, and sampling. Structural learning involves estimating the dependency relationships between variables from data, while parameter learning estimates conditional probability distributions. By importing bnlearn and reading data, it is easy to create Bayesian network structure diagrams. The specific operation is as follows:

Step 1. Import bnlearn and pandas libraries
Step 2. Read and analyze data
Step 3. Conduct structural learning
Step 4. Parameter learning
Step 5. Reasoning
Step 6. Draw a Directed Acyclic Graph (DAG)

4.2 Program Implementation Code

Based on the above algorithm, the specific implementation code is as follows:

```
Import bnlearn as bn
Import pandas as pd
df=pd.read_csv("GWRYSB3.csv")
model=bn.structure_learning.fit(df)          #Structural learning
G=bn.plot(model)                                      #Draw a picture
model1=bn.parameter_learning.fit(df)      #Parameter learning
G1=bn.plot(model1)                                 #Reasoning
model2=bn.inference.fit(df)
G3=bn.plot(model2)                                 #Draw a picture
model3=bn.import_DAG(df)                   # load buf file
G3=bn.plot(model3)                                 #Draw a picture
```

4.3 *Results of Creating a Cause and Effect Diagram for Drug Crimes Based on Python*

Using Python's bnlearn and pandas libraries, guided by Bayesian networks and causal relationship theory, based on data with 21 field attributes, a directed acyclic graph (DAG) containing 12 field attributes was obtained after analysis, as shown in Fig. 1.

Based on this Bayesian network DAG graph, the following causal diagram is obtained, as shown in Fig. 2a–d:

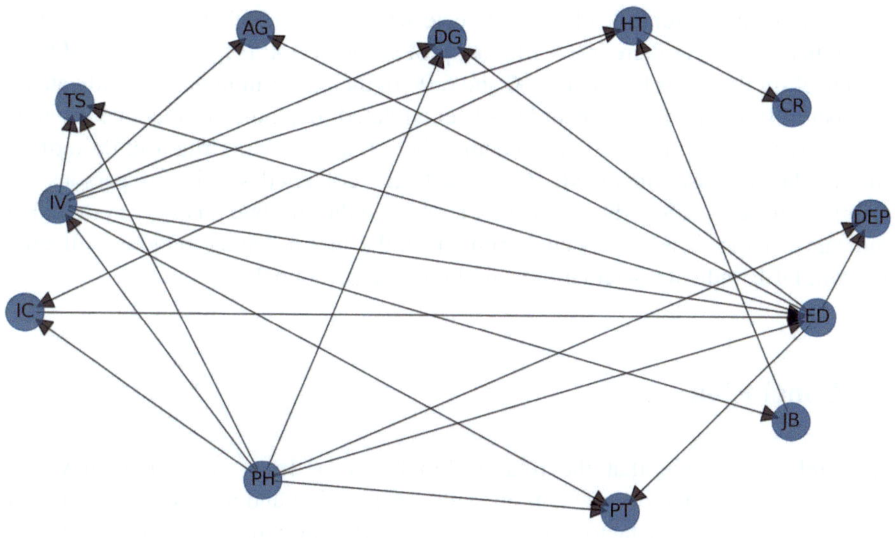

Fig. 1 DAG graph based on Python

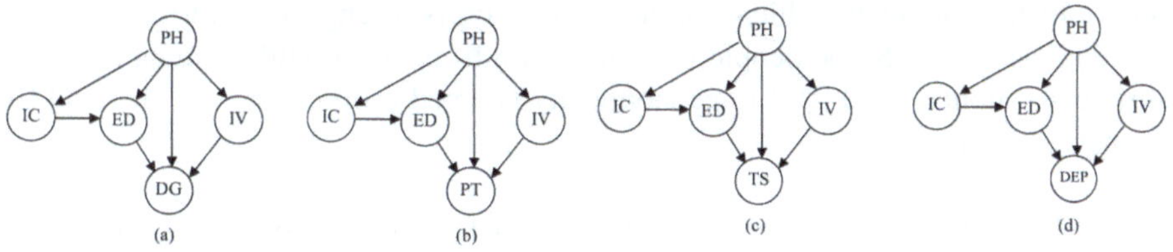

Fig. 2 Bayesian network causality diagram based on drug crimes

4.4 Causal Relationship Based on Drug Crime Data

According to the causal relationship diagram shown in Fig. 2, corresponding causal relationships can be obtained, such as:

PH → DG, PH → PT, PH → TS, PH → DEP, PH → ED → DG, PH → ED → PT, PH → ED → TS, PH → ED → DEP, PH → IC → ED → DG, PH → IC → ED → PT, PH → IC → ED → TS, PH → IC → ED → DEP, and so on.

Based on the causal relationship obtained from the data above, it is easy to infer the corresponding results under given conditions (causes).

Note: The code symbols for the causal relationships above are shown:

ED, indicate cultural level; AG, indicate whether the age is between 18 and 40; JB, indicate whether there is a legitimate profession or not; HT, the number of times the check-in time was between 0:00 and 6:00 in the past 3 months; CR, indicate the number of times a criminal record has been violated; IC, indicates that there is an IC card storage record; DG, indicate whether there is a history of drug use; IV, indicate the number of cases involved; PT, indicate the collected suspicious persons and criminal suspects; TS, indicate whether it exists in the sub-bureau query system; PH, indicate whether it is a suspected criminal in the mobile phone number collection system; DEP, indicate whether they are high-risk personnel.

5 Conclusion

Research has shown that the relationship between drugs and crime may vary depending on risk factors, deviant intensity, progression and maintenance stages, as well as environmental factors. Combining expert experience, A person's lifestyle is gradually established through repeated actions and the meanings attached to these actions, as well as numerous interactions between individuals, environments, and time environments, which encourage drug users to engage in various types of criminal activities. Some people may transition from successful criminal activities (crime → money → drugs) to economically coerced crime (drugs → demand for money → crime).

Acknowledgments This work partially fulfills research program and has received funding from Key Laboratory of Intelligent Drug Control (Project Number:ZHJDNB202401); funding from a key research project at the university level by Yunnan Police College (Project Number: 23A002).

References

1. Speckart, G., Anglin, M.D.: Narcotics and crime: an analysis of existing evidence for a causal relationship. Behav. Sei. Law. **3**, 259–283 (1985)

2. Weissman, J.C.: Understanding the drugs and crime connection. In: Weissman, J.C., DuPont, R.L. (eds.) Criminal Justice and Drugs: the Unresolved Connection, National University Publications, Port Washington, N.Y (1982)
3. Pearl, J.: Theoretical impediments to machine learning with seven sparks from the causal revolution. arXiv preprint arXiv:1801.04016 (2018)
4. Dehejia, R.H., Wahba, S.: Causal effects in nonexperimental studies: reevaluating the evaluation of training programs. J. Am. Stat. Assoc. **94**(448), 1053–1062 (1999)
5. Mani, S., Cooper, G.F.: Causal discovery from medical textual data. In: Proceedings of the AMIA Symposium, p. 542 (2000)
6. Imbens, G.W.: Nonparametric estimation of average treatment effects under exogeneity: a review. Rev. Econ. Stat. **86**(1), 4–29 (2004)
7. Pearl, J.: Causal inference in statistics: an overview *. Stat. Surv. **3**(2009), 96–146 (2009)
8. Gelman, A.: Causality and statistical learning. AJS. **117**(3), 955–966 (2011)

Resource Allocation and Value of Service Optimization for Integrated Sensing and Communication-Enabled Internet of Vehicles

Limin Jiang and Xinjie Yang

1 Introduction

With the commercial rollout of 5G mobile communication systems and the rapid development of Internet of Things (IoT) and Artificial Intelligence (AI) technologies, Internet of Vehicle (IoV) networks are rapidly emerging and gradually changing the way people travel [1]. Emerging applications of IoV require high-precision sensing capabilities and high-speed data communication services under complex wireless conditions [2]. Therefore, base station (BS) needs to balance the sensing and communication performance of vehicular users under the condition of limited wireless resources. Researchers have proposed various resource allocation strategies and evaluation metrics [3, 4]. Most studies aim to maximize sensing performance while constraining the proportion of communication data. In [3], Zhang et al. proposed a Joint Sensing and Communication Integrated System (JSCIS) that optimizes real-time sharing of inter-vehicular sensing data through dynamically configuring frame structure. The objective is to maximize Mutual Information (MI) while constraining the communication channel capacity for multiple connected automated vehicles. Some studies also aim to maximize data transmission rates while ensuring that achievable sensing performance reaches an acceptable level, as in [4]. Additionally, some research introduces mechanisms aimed at simultaneously optimizing sensing and communication performance.

L. Jiang
Faculty of Electrical Engineering and Computer Sciences, Ningbo University, Ningbo, China

X. Yang (✉)
Faculty of Electrical Engineering and Computer Sciences, Ningbo University, Ningbo, China

Ningbo ArrowDeep Internet Technology Co., Ltd., Ningbo, China
e-mail: yangxinjie@nbu.edu.cn

X. Jia et al. (eds.), *2nd International Conference on Big Data, IoT, and Cloud Computing*, EAI/Springer Innovations in Communication and Computing,
https://doi.org/10.1007/978-3-031-93570-1_18

The limited wireless resources in telematics can affect the reliability and efficiency of data transmission, which in turn affects the quality of service. ISAC technology offers a potential solution for efficient resource management by integrating sensing and communication functions that can significantly improve spectrum utilization and system performance to better meet the stringent requirements of the Internet of Vehicles. By sharing spectrum resources between sensing and communication, ISAC reduces the need for dedicated frequency bands, thereby enhancing spectrum efficiency. Additionally, the ISAC system minimizes interference and optimizes resource allocation, ensuring more reliable and efficient data transmission. In the communication and sensing process, hardware platforms and radio resources need to be shared, which leads to resource competition between sensing and communication. Therefore, there is an urgent need for a comprehensive evaluation metric to achieve resource balance between sensing and communication based on user requirements [5]. In this chapter, we aim to simultaneously satisfy vehicle users' demands for communication and sensing services under limited wireless resources. The main contributions of this chapter are as follows:

- A scenario consisting of multiple BSs and multiple vehicular users is studied, where vehicular users communicate with the BS while sensing target vehicles ahead. To evaluate the performance of vehicular users, a new metric of weighted Value of Service (VoS) is introduced to balance VoS of both communication and sensing. The goal of the optimization is to maximize weighted VoS.
- An algorithm is proposed to jointly optimize user association and subchannel allocation, transforming the original NP-hard optimization problem into two solvable subproblems.

The remaining of this chapter is organized as follows. In Sect. 2, we describe the system model and problem formulation. In Sect. 3, the algorithm to solve the problem is explained. In Sect. 4, the simulation results are analyzed and compared with benchmark algorithms, and finally, Sect. 5 concludes the work.

2 System Model and Problem Formulation

2.1 ISAC-Enabled IoV System Model

As shown in Fig. 1, consider an IoV system consisting of a one-way highway with two lanes, where vehicles are traveling unidirectionally from right to left. Along the highway, a set of BSs $\mathcal{M} = \{1, 2, \ldots, m, \ldots, M\}$ is located. All BSs share a set of orthogonal subchannels $\mathcal{V} = \{1, 2, \ldots, v, \ldots, V\}$, and each subchannel has a bandwidth of B. The M subchannels are allocated exclusively to all BSs, and $v_m \in \mathcal{V}$ represents the number of subchannels available to BS m. Each Vehicle User Equipment (VUE) is assigned a subchannel that is used for performing both sensing

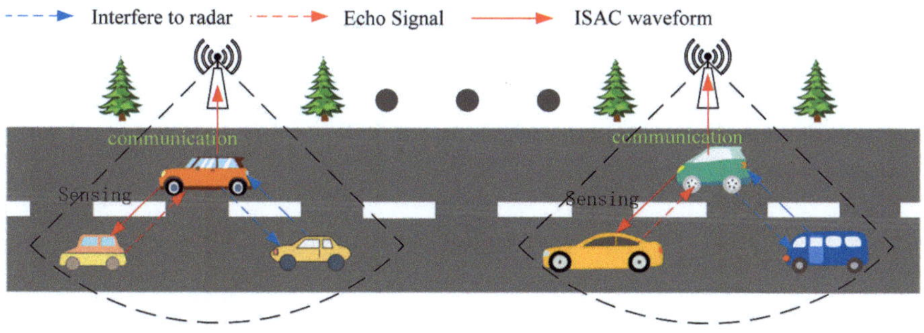

Fig. 1 System model

and communication. It is assumed that the m-th BS provides services to I_m VUEs within its service area.

The ISAC beams integrate communication and sensing functions via technologies known as beam-forming and joint optimization, facilitating efficient data transmission for communication users and target detection and localization through beam scanning. ISAC technologies can therefore contribute to substantial enhancement in resource utilization efficiency and have found extensive applications in domains such as intelligent transportation, industrial automation, and smart city development. The set of VUEs is denoted as $\mathcal{I} = \{1, 2, \ldots, i, \ldots, I\}$. It is assumed that all VUEs are equipped with ISAC technology. VUEs can use ISAC signals for sensing the vehicles ahead and for Vehicle-to-Infrastructure (V2I) communication. Considering that each VUE is associated with computing tasks, VUEs need to sense the surrounding vehicles and environment through various sensors (including radar) and upload the sensed data via V2I links to the service BS. The sensed data includes the position, speed, direction, of surrounding vehicles, as well as road and traffic flow information.

When a VUE performs ISAC tasks, the VoS is established for communication and sensing separately. Finally, an integrated VoS for the system is derived by considering the weighting factors for sensing and communication. The optimization goal is to maximize the system's VoS.

2.2 *VoS of Communication*

In the process of communication, each VUE sends sensed data to its serving BS using its allocated subchannel; therefore, it is necessary to consider the matching of VUEs to BSs and also to subchannels. Defining the binary variable $y_{i,m}$ to represent the association between VUE $i \in \mathcal{I}$ and BS $m \in \mathcal{M}$, then, the matrix $Y = [y_{i,m}]_{I\times M}$ is defined as the matching matrix

$$y_{i,m} = \begin{cases} 1, & \text{if VUE } i \text{ is served by BS } m \\ 0, & \text{otherwise} \end{cases} \tag{1}$$

Similarly, defining the binary variable $x_{v,i}$ to indicate that the subchannel $v \in \mathcal{V}$ is allocated to VUE $i \in \mathcal{I}$,

$$x_{v,i} = \begin{cases} 1, & \text{if subchannel } v \text{ is allocated to VUE } i \\ 0, & \text{otherwise} \end{cases} \tag{2}$$

Then the matrix $X = [x_{v,i}]_{V \times I}$ is defined as the allocation matrix.

The system considered in this chapter is an IoV system based on Orthogonal Frequency Division Multiplexing Access (OFDMA) technology. For VUE $i \in \mathcal{I}$, its achievable transmission rate, with BS $m \in \mathcal{M}$ and subchannel $v \in \mathcal{V}$ combination, according to Shannon's theory, is

$$R_{i,m}^{v} = B_{i,m}^{v} \log_2 \left(1 + \text{SINR}_{i,m}^{v}\right), \forall i \in \mathcal{I}, m \in \mathcal{M}, v \in \mathcal{V} \tag{3}$$

where $B_{i,m}^{v}$ represents the bandwidth of subchannel v. The Signal to Interference plus Noise Ratio (SINR) at BS m when receiving signal from VUE i is given by

$$\text{SINR}_{i,m}^{v} = \frac{y_{i,m} x_{v,i} P_i |\mathcal{H}_{i,m}|^2 d_{i,m}^{-\alpha}}{\sigma_c^2 + \sum_{i' \in \mathcal{I} \setminus \{i\}} x_{v,i'} h_{i',m}^{v} P_{i'}} \tag{4}$$

where P_i is the transmission power of VUE i for data transmission, $\mathcal{H}_{i,m}$ represents the small-scale Rayleigh fading coefficient between BS m and VUE i, following a complex Gaussian distribution $CN(0, 1)$, $d_{i,m}$ denotes the distance between the VUE i and the BS m, α is the large-scale fading factor, σ_c^2 is the power of additive white Gaussian background noise, and $h_{i',m}^{v}$ and $P_{i'}$ are the channel gain and transmission power of interferer VUE i'.

The achievable transmission rate for VUE i is

$$R_i = \sum_{v \in V} \sum_{m \in \mathcal{M}} R_{i,m}^{v} \tag{5}$$

To assess the performance of the communication process of VUE i, we define the VoS of communication specific to VUEs as follows [6]:

$$VC_i = \begin{cases} \frac{R_i}{R_i^{req}}, & if \quad R_i < R_i^{req} \\ 1, & if \quad R_i \geq R_i^{req} \end{cases} \tag{6}$$

where R_i^{req} is the maximum transmission rate required by VUE i.

2.3 VoS of Sensing

In the sensing process, VUE $i \in \mathcal{I}$ radiates an integrated OFDM waveform through its dual-function transmitter, using the same subchannel for communication. The integrated sensing and communication waveform can be used for radar sensing of vehicles ahead. On subchannel $v \in \mathcal{V}$, the waveform used by VUE $i \in \mathcal{I}$ consists of K consecutive integrated symbols, which can be described as follows [7]:

$$s_i^v(t) = e^{j2\pi f_c t} \sum_{\tau=0}^{K-1} a_i^v c_i^{\tau,v} e^{j2\pi i \Delta f (t-\tau T_0)} \cdot \text{rect}\left[\frac{t-\tau T_0}{T_0}\right] \tag{7}$$

where $T = 1/\Delta f$ is the duration of the basic symbol of the OFDM symbol, f_c is the center frequency of subchannel v, and T_0 is the duration of a single OFDM symbol with a cyclic prefix (CP), i.e., the time duration of a complete OFDM symbol. Specifically, $T_0 = T + T_g$, where T_g is the CP duration. a_i^v represents the amplitude of the waveform of VUE i on subchannel v, $c_i^{\tau,v}$ represents the phase encoding of the τ-th modulation symbol of VUE i on subchannel v, rect$[x]$ is a rectangular function, which value is 1 when $0 \le x \le 1$, and 0 otherwise.

Assuming the impulse response $g_i^v(t)$ of VUE i on subchannel v is a Gaussian random process, then, the reflected signal received by VUE $i \in \mathcal{I}$ on subchannel $v \in \mathcal{V}$ can be expressed as

$$y_i(t) = \int_{-\infty}^{+\infty} g_i^v(\psi) s_i^v(t-\psi) d\psi + z_i(t) \tag{8}$$

where the received echo signal is affected by additive white Gaussian noise $z_i(t)$. In the assessment of radar target detection performance, the conditional MI can be utilized to evaluate the accuracy of radar target's impulse response. During the sensing process, since the same subchannel is shared with communication, only the matrix X needs to be considered, which is defined in Eq. (2).

The conditional MI for the VUE i on subchannel v is given by [8]

$$MI_i^v = I(y_i(t); g_i^v(t) \mid s_i^v(t)) = \frac{1}{2} T_p \Delta f \log_2(1 + \Gamma_i^v) \tag{9}$$

where $T_p = K T_0$ represents the total duration of the OFDM symbols. Over T_p , the average MI that VUE i can achieve is [9]

$$\overline{MI}_i = \sum_{v \in \mathcal{V}} \frac{MI_i^v}{T_p} \tag{10}$$

The SINR for the echo signal received by VUE i in sensing is

$$\Gamma_i^v = \frac{x_{v,i} P_i K T_0^2 |G_i^v(f)|^2}{\sigma_s^2 + \sum_{i' \in \mathcal{I} \setminus \{i\}} x_{v,i'} h_{i',i}^v P_{i'}} \tag{11}$$

where $G_i^v(f)$ is the Fourier transform of $g_i^v(t)$, the interference comes from other interfering radar signals from VUE $i' \in \mathcal{I} \setminus \{i\}$.

Similar to the VoS of communication, the VoS of sensing can be defined as

$$VS_i = \begin{cases} \frac{\overline{MI}_i}{MI_i^{req}}, & if \quad \overline{MI}_i < MI_i^{req} \\ 1, & if \quad \overline{MI}_i \geq MI_i^{req} \end{cases} \tag{12}$$

where MI_i^{req} is the maximum MI required by VUE i.

2.4 Problem Formulation

The VoS of each VUE is designed as the weighted sum of VoS of communication and VoS of sensing, and the system's VoS is the average of all VUEs' VoS

$$\mathrm{VoS}(Y, X) = \frac{1}{I} \sum_{i \in \mathcal{I}} (\alpha_i VC_i + \beta_i VS_i) \tag{13}$$

where $\alpha_i > 0$, $\beta_i > 0$, and $\alpha_i + \beta_i = 1$, and α_i and β_i represent the communication capability and sensing accuracy of VUE i, respectively.

Given the limited bandwidth resources of the frequency band, it is crucial to balance the performance of communication and sensing to avoid detrimental competition. Accordingly, an optimization problem is formulated to maximize the system's VoS by jointly optimizing the association and allocation variables, as follows:

$$\begin{aligned} \mathcal{P}1 : & \max_{Y,X} \mathrm{VoS}(Y, X) \\ \text{s.t.} \quad & C1 : MI_i \geq MI_i^{\min}, \quad i \in \mathcal{I}, \\ & C2 : R_i \geq R_i^{\min}, \quad i \in \mathcal{I}, \\ & C3 : x_{v,i} \in \{0, 1\}, \quad i \in \mathcal{I}, \quad v \in \mathcal{V}, \\ & C4 : \sum_{v \in \mathcal{V}} x_{v,i} = 1, \quad i \in \mathcal{I}, \\ & C5 : y_{i,m} \in \{0, 1\}, \quad i \in \mathcal{I}, \quad m \in \mathcal{M}, \\ & C6 : \sum_{m \in \mathcal{M}} y_{i,m} = 1, \quad i \in \mathcal{I}, \\ & C7 : \sum_{i \in \mathcal{I}} y_{i,m} \leq I_m, \quad m \in \mathcal{M} \end{aligned} \tag{14}$$

Constraint C1 ensures the minimum sensing performance $MI_i^{\min}$ for each vehicle. Constraint C2 ensures that each vehicle has a minimum communication performance $R_i^{\min}$. Constraint C3 indicates that each element of the matching matrix is a binary variable. Constraint C4 specifies that each VUE can only be allocated with one V2I subchannel. Constraint C5 indicates that each element of the allocation matrix is a binary variable. Constraint C6 specifies that each VUE can only be served by one BS. Constraint C7 states that for BS m, a maximum of I_m VUEs can be served. In $\mathcal{P}1$, the objective function is nonlinear due to the presence of binary variables. In addition, the existence of multiplicative relationships among other variables, along with constraints C3 and C5, renders the feasible set non-convex. As a result, the problem falls into a Mixed-Integer Nonlinear Programming (MINLP) category, which is non-convex and NP-hard [9]. Therefore, solving this problem is extremely challenging.

3 Proposed Scheme

Due to the complexity of the original problem $\mathcal{P}1$, it can be decoupled into two subproblems, namely the problem of association between VUEs and BSs and the problem of subchannel allocation. Matching theory is an effective approach to solving such complex optimization problems by transforming the problem into an efficient pairwise model. It can be seen that both subproblems involve finding M-to-N point association; therefore, the matching theory can be applied to solve them with alternating iterations. So both subproblems can be solved using the matching theory, and alternative loops are used to allow the solution to converge.

3.1 VUEs and BS Association Problem

Given a known subchannel allocation matrix X_0, the problem becomes maximizing $\text{VoS}(X_0, Y)$, with optimization space being the matching matrix Y. Y satisfies that each VUE can be served by one BS, and each BS can accept up to I_m VUEs. The specific objective function and constraints can be transformed as follows:

$$\begin{aligned} \mathcal{P}1.1 : \max_{Y} \text{VoS}(X_0, Y) &= \frac{1}{I} \sum_{i \in \mathcal{I}} (\alpha_i VC_i + \beta_i VS_i) \\ \text{s.t.} \quad & C2, C5, C6, C7 \end{aligned} \tag{15}$$

Defining E_m as the set of VUEs served by the m-th BS, $\mathcal{P}1.1$ can be rewritten as

$$
\begin{aligned}
&\mathcal{P}1.1.1: \max_{E_m} \frac{1}{I} \sum_{m \in \mathcal{M}} O_m(E_m) \\
&\text{s.t.} \quad C2: R_i \geq R_i^{\min}, \quad i \in \mathcal{I}, \\
&\qquad C5\text{-}1: \bigcup_{m \in \mathcal{M}} E_m = \mathcal{I}, \\
&\qquad C6\text{-}1: E_{m_1} \cap E_{m_2} = \emptyset, \quad \forall m_1 \neq m_2, \\
&\qquad C7\text{-}1: |E_m| \leq I_m,
\end{aligned} \tag{16}
$$

$$
O_m(E_m) \triangleq \alpha_i \sum_{i \in E_m, v \in \mathcal{V}} \frac{R_{i,m}^v}{R_i^{req}} + \beta_i \sum_{i \in E_m, v \in \mathcal{V}} \frac{\overline{MI}_i}{MI_i^{req}}
$$

$\mathcal{P}1.1.1$ is calculated from the BS perspective, while $\mathcal{P}1.1$ is calculated from the VUE's perspective. Therefore, the original user matching problem is transformed from the optimization of Y to the optimization of E_m. The goal of $\mathcal{P}1.1.1$ is to divide all I VUEs into M sets, corresponding to the VUEs connected to M BSs. The objective function is the same, i.e., to maximize the overall system VoS. To solve this problem, the matching theory is used to transform the many-to-many matching problem into a many-to-one matching problem. We define the set of M BSs as $\mathcal{B}_M = \{B_1, \ldots, B_m, \ldots, B_M\}$, where B_m represents the m-th BS. Similarly, we define the set of I VUEs as $\mathcal{V}_I = \{V_1, \ldots, V_i, \ldots, V_I\}$, where V_i represents the i-th VUE. Next, we introduce the matching theory based on this definition [10].

Definition 1 (Many-to-One Matching Model) Let V_i be the i-th VUE and B_m be the m-th BS, and the association between the VUE set and the BS set involves a many-to-one matching model. Therefore, the process of pairing of VUE with BS can be defined with a matching function as follows:

$$
\begin{aligned}
M(x) &\in \{V_1, \ldots, V_i, \ldots, V_I\} \cup \{B_1, \ldots, B_m, \ldots, B_M\}, \\
x &\in \{V_1, \ldots, V_i, \ldots, V_I\} \cup \{B_1, \ldots, B_m, \ldots, B_M\}
\end{aligned} \tag{17}
$$

This model is defined as follows:

- $M(V_i) \subseteq \{B_1, \ldots, B_m, \ldots, B_M\}$, and $|M(V_i)| = 1$.
- $M(B_m) \subseteq \{V_1, \ldots, V_i, \ldots, V_I\}$, and $|M(B_m)| \leq I_m$.
- $M(V_i) = B_m$, if and only if $V_i \in M(B_m)$.

Definition 2 (Preference Degree) For VUE V_i, the preference degree for any BS B_m is defined as the performance gain brought to the BS when VUE V_i is served by BS B_m, that is,

$$
G_{i,m} = O_m(E'_m \cup i) - O_m(E'_m) \tag{18}
$$

where E'_m is the set of VUEs other than V_i that are connected to BS B_m. Based on this definition, for VUE i, V_i prefers to connect to BS B_{m_1} if and only if $G_{i,m_1} > G_{i,m_2} > \cdots$. Similarly, for BS m, B_m prefers to allow V_i to connect if and only if $G_{i_1,m} > G_{i_2,m} > \cdots$.

3.2 Subchannel Allocation Problem

Given a known VUE-BS matching matrix Y_0, this subproblem is to solve for X. Note that the constraints containing variable X come from C1, C2, C3, and C4, as shown below:

$$\begin{aligned} \mathcal{P}1.2 : \max_{X} \text{VoS}(X, Y_0) = \frac{1}{I} \sum_{i \in \mathcal{I}} (\alpha_i VC_i + \beta_i VS_i) \\ \text{s.t.} \quad C1 \sim C4 \end{aligned} \tag{19}$$

From subchannel's perspective, define A_v as the set of VUEs served by the v-th subchannel. Then, problem $\mathcal{P}1.2$ is equivalent to

$$\begin{aligned} &\mathcal{P}1.2.1 : \max_{A_v} \frac{1}{I} \sum_{v \in \mathcal{V}} O_v(A_v) \\ &\text{s.t.} \quad C1 \sim C2, \\ &\qquad C3\text{-}1 : \bigcup_{v \in \mathcal{V}} A_v = \mathcal{V}, \\ &\qquad C4\text{-}1 : A_{v_1} \cap A_{v_2} = \emptyset, \quad \forall v_1 \neq v_2, \\ &O(A_v) \triangleq \alpha_i \sum_{i \in A_v, m \in \mathcal{M}} \frac{R_{i,m}^v}{R_i^{req}} + \beta_i \sum_{i \in A_v} \frac{\overline{MI}_i}{MI_i^{req}} \end{aligned} \tag{20}$$

Therefore, the original channel allocation problem is transformed from the optimization of X to the optimization of A_v. Like the first subproblem, the matching theory can be applied to solve this problem.

3.3 Cooperative User Association and Channel Allocation Iterative Optimization

The original problem $\mathcal{P}1$ is decoupled into two subproblems, and each of them is solved separately, with iterative loops. During each iteration, the optimal solutions of the subproblems are calculated, and the optimization process is terminated after reaching the maximum number of iterations, or the difference of consecutive results is smaller than a defined threshold. This scheme is named MUA (Matching algorithm of joint BS-User Association and subchannel allocation), as shown in Algorithm 1. Since the maximum number of iterations is k, I VUEs, M BSs, and V subchannels, the algorithmic complexity of the problem is $O\big(k \cdot I(M+V+\log(I))\big)$.

Algorithm 1: Matching algorithm of joint BS-User Association and subchannel allocation

Input: Minimum transmission rate $R_i^{\min}$, minimum MI $MI_i^{\min}$, maximum number of iterations $k_{\max}$.
Output: Optimized user-to-BS association variable Y and subchannel allocation variable X.
Initialize $Y^{(0)}$ and $X^{(0)}$. Set $k = 0$.
repeat
  Step 1: BS-User Association;
  Initialize each BS m with $E_m = \emptyset$. Set $\mathcal{B}^a = \mathcal{B}_M, \mathcal{U}^U = \mathcal{V}_I, t = 1$.
  while $t \leq T$ **do**
    $m^* = \text{argmax}_m\, G_{i,m}$, $i^* = \text{argmax}_i\, G_{i,m}^t$.
    if $\mathcal{U}^U = \emptyset$ ***or*** $\mathcal{B}^a = \emptyset$ **then**
    else
      $t = t + 1$.
    end if
  end while
  Step 2: subchannel Allocation;
  Initialize each subchannel v with $A_i = \emptyset$. Set $\mathcal{V}, \mathcal{U}, t = 1$.
  while $t \leq T$ **do**
    $v^* = \text{argmax}_v\, G_{i,v}$, $i^* = \text{argmax}_i\, G_{i,v}^t$.
    if $\mathcal{U} = \emptyset$ ***or*** $\mathcal{V} = \emptyset$ **then**
    else
      $t = t + 1$.
    end if
  end while
  Step 3: Evaluate Objective Function;
  Calculate $\Theta(k) = VoS(Y^{(k+1)}, X^{(k+1)})$.
  $k = k + 1$.
until $k > k_{max}$;
return *Optimized variable Y^* and X^*.*

4 Simulation Results

For the communication model, it is assumed that the small-scale fading channel gain follows Rayleigh fading with a unit mean. The uplink path loss is determined by the large-scale loss factor $\alpha = 2$ and the distance $d_{i,m}$ between the VUE and its associated BS. For the sensing model, it is assumed that the target frequency

Table 1 Simulation parameters and settings

Parameter	Value	Parameter	Value
P_i	200 mW	B	1 MHz
T_0	5×10^{-6} s	K	10
$R_{\min}$	0.1 Mbps	$MI_{\min}$	40 bit
σ_c^2	−174 dBm/Hz	σ_s^2	−74 dBm/Hz

response follows a standard normal distribution. Other key parameters and their settings are shown in Table 1. To thoroughly assess the performance of the proposed MUA scheme, the following often referred to three schemes are considered as benchmarks.

(1) MUA-FP (Matching algorithm for BS and User Association-Fractional Programming) [11]: This method uses matching theory to optimize the first subproblem, i.e., the association problem between VUEs and BSs, while using the fractional programming to solve the second subproblem, i.e., subchannel allocation problem.
(2) IMUA-FP (Improved Matching algorithm for BS and User Association-Fractional Programming): Comparing to MUA-FP, this scheme introduces the Hungarian algorithm [12] to directly solve the first subproblem based on the original matching theory. The number of iterations in the simulations is significantly reduced however, with loss of performance. The second subproblem is solved by fractional programming, similarly.
(3) GA (Genetic Algorithm): Encoding the association matrix Y and the subchannel allocation matrix X into the chromosome and defining the population and fitness function. The solution can be intelligently searched, employing natural selection, crossover, mutation, and exploration of genes. In the simulations, the population size is set as 100, the maximum number of evolutions as 300, and the crossover probability and the mutation probability as 0.0001 and 0.1, respectively.

The simulations are performed to assess the performance of an ISAC-enabled IoV system with the proposed scheme under varying numbers of VUEs (service requests). In the scenario varying communication and sensing tasks, the resource allocation has to be robust to meet different traffic demand. In the simulations, the number of VUEs varies between 3 and 30. Figure 2 shows the system VoS, VoS of communication, and VoS of sensing, respectively, when subchannel number is fixed to 15. It is demonstrated that when VUE number increases, all the VoS decreases, and the VoS values of the proposed MUA scheme outperform all three benchmark algorithms. When the number of VUEs is less than 7, all four algorithms perform well in terms of VoS. However, with a larger VUE number, all algorithms except the proposed MUA exhibit a significant decrease in VoS. When the number of VUEs exceeds 24, the VoS values of MUA-FP, IMUA-FP, and GA all drop below 0.7, while the VoS of the proposed MUA remains above 0.85. This is because MUA comprehensively considers the competition between VUEs using the same

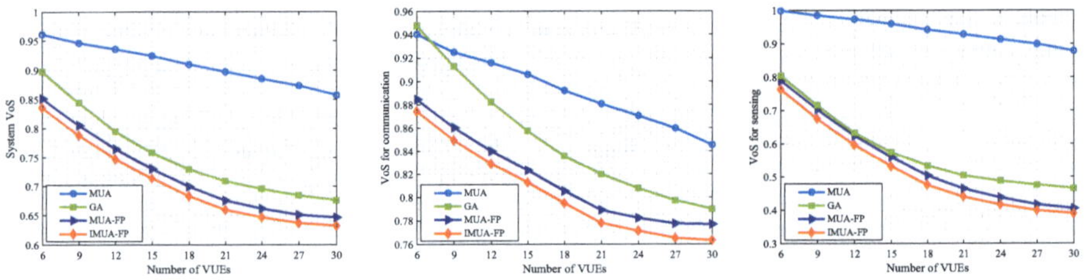

(a) System VoS (b) VoS of communication (c) VoS of sensing

Fig. 2 (**a**), (**b**), and (**c**) under different numbers of VUEs

resource (BS/subchannel), whereas the other three algorithms do not. GA achieves relatively high VoS due to its iterative evolution, which optimizes variable selection and achieves a result close to that of the exhaustive search. MUA-FP is slightly better than IMUA-FP in performance, as it prioritizes higher VUE-BS pairs with higher preference degrees when associating VUEs with BSs in every iteration.

5 Conclusion

This chapter investigates an ISAC-enabled IoV system comprising multiple RSUs and VUEs. To maximize system VoS, it jointly optimizes VUE-BS associations and subchannel allocation. A novel MUA scheme based on matching theory is proposed to address this optimization challenge. For performance benchmarking, three widely recognized ISAC resource allocation schemes are selected. Comprehensive simulations demonstrate that the proposed MUA scheme achieves significantly higher VoS across varying levels of network complexity.

References

1. Zhang, Q., Sun, H., Gao, X., Wang, X., Feng, Z.: Time-division ISAC enabled connected automated vehicles cooperation algorithm design and performance evaluation. IEEE J. Sel. Areas Commun. **40**(7), 2206–2218 (2022)
2. Wang, W., Zhang, S.Z., Li, X.F., et al.: 5G-advanced technology and applications. ZTE Technol. J. **30**(4), 69–76 (2024)
3. Zhang, Q., Wang, X., Li, Z., Wei, Z.: Design and performance evaluation of joint sensing and communication integrated system for 5G mmWave enabled CAVs. IEEE J. Sel. Top Signal Process. **15**(6), 1500–1514 (2021)
4. Xu, D., Yu, X., Ng, D.W.K., Schmeink, A., Schober, R.: Robust and secure resource allocation for ISAC systems: a novel optimization framework for variable-length snapshots. IEEE Trans. Commun. **70**(12), 8196–8214 (2022)
5. Li, B., Wang, X., Xin, Y., Au, E.: Value of service maximization in integrated localization and communication system through joint resource allocation. IEEE Trans. Commun. **71**(8), 4957–4971 (2023)

6. Li, B., Wang, X., Fang, F.: Maximizing the value of service provisioning in multi-user ISAC systems through fairness guaranteed collaborative resource allocation. IEEE J. Sel. Areas Commun. **42**(9), 2243–2258 (2024)
7. Liu, Y., Liao, G., Xu, J., Yang, Z., Zhang, Y.: Adaptive OFDM integrated radar and communications waveform design based on information theory. IEEE Commun. Lett. **21**(10), 2174–2177 (2017)
8. Zhao, L., Wu, D., Zhou, L., Qian, Y.: Radio resource allocation for integrated sensing, communication, and computation networks. IEEE Trans. Wirel. Commun. **21**(10), 8675– 8687 (2022)
9. Yang, J., Fu, Y., Li, C., Yuan, X.: Joint optimization scheme for user association and resource allocation in internet of vehicles. In: 2023 IEEE 98th Vehicular Technology Conference (VTC2023-Fall), Hong Kong, pp. 1–5 (2023)
10. LeAnh, T., et al.: Matching theory for distributed user association and resource allocation in cognitive femtocell networks. IEEE Trans. Vehic. Technol. **66**(9), 8413–8428 (2017)
11. Shen, K., Yu, W.: Fractional programming for communication systems—Part I: power control and beamforming. IEEE Trans. Signal Process. **66**(10), 2616–2630 (2018)
12. Yin, Z., Sun, Y., Gu, M., Hu, H., Ren, Z., Jin, Z.: Research on deep sea search and rescue based on Markov Chain Monte Carlo method and hungarian algorithm. In: 2024 IEEE 2nd International Conference on Image Processing and Computer Applications (ICIPCA), Shenyang, pp. 1786–1790 (2024)

A Novel Adaptive Access Mechanism for Cellular V2X Networks

Yijing Li, Xinjie Yang, and Pei Xiao

1 Introduction

With the rapid development of intelligent transport systems toward automated driving, telematics communication as a key technology has received extensive attention from both academia and industry [1]. The traditional Dedicated Short-Range Communication (DSRC) is the key technology to realize Vehicle-to-Vehicle (V2V) communication. Although DSRC can meet the basic communication requirements such as delay, Quality of Service (QoS), preventing collisions, etc., it cannot meet the strict reliability requirements in the presence of high network load [2, 3]. Therefore, the Third Generation Partnership Project (3GPP) introduced the first Cellular Vehicle-to-Everything (C-V2X) technology in Release 14, named as Long-Term Evolution-V2X (LTE-V2X) [4, 5]. LTE-V2X consists of two modes named Mode 3 and Mode 4. To meet the stringent requirements for autonomous driving and to improve the flexibility and stability of V2X communications, the 3GPP has advanced the development of C-V2X to New Radio-V2X (NR-V2X) for 5G (the Fifth Generation).

NR-V2X supports two modes, namely Mode 1 and Mode 2 for radio resource allocation, and these two modes are improvements of Mode 3 and Mode 4,

Y. Li
Faculty of Electrical Engineering and Computer Sciences, Ningbo University, Ningbo, China

X. Yang (✉)
Faculty of Electrical Engineering and Computer Sciences, Ningbo University, Ningbo, China

Ningbo Arrowdeep Internet Technology Co. Ltd, Ningbo, China
e-mail: yangxinjie@nbu.edu.cn

P. Xiao
ICS, University of Surrey, Guildford, UK

X. Jia et al. (eds.), *2nd International Conference on Big Data, IoT, and Cloud Computing*, EAI/Springer Innovations in Communication and Computing,
https://doi.org/10.1007/978-3-031-93570-1_19

respectively. Within Mode 1, resources are distributed by the base station in a centralized manner, whereas in Mode 2, resources are allocated and also reserved in a distributed manner, benefiting vehicles outside the base station's coverage. A sensing-based Semi-Persistent Scheduling (SPS) scheme was initially presented in Mode 4, and it has been improved in Mode 2 to reduce packet conflicts during resource reservation by introducing an NR-numerologies and reevaluation mechanism [6]. These features will provide more flexible, more reliable, and higher throughput V2X services for advanced driving applications [7].

However, as traffic density increases, channel contention, collision, and retransmission probabilities increase. This results in periodic beacon messages occupying too much of the control channel, which leads to channel congestion and affects the Packet Delivery Rate (PDR) [8]. To cope with this challenge, this chapter proposes a novel ADaptive Access Mechanism to improve 5G NR-V2X Mode 2, named as ADAM-M2. It enhances the existing SPS scheme and dynamically adjusts the Modulating and Coding Schemes (MCSs) and sub-channelization schemes for Vehicle User Equipment (VUE) under different communication conditions to ensure the efficient transmission of information. The Channel Busy Ratio (CBR) is utilized as a selection criterion for selecting MCS and the two sub-channelization schemes. Moreover, the PDR is used as the main metric for evaluating the scheme performance.

2 5G NR-V2X Mode 2 Physical Layer

5G NR-V2X uses Orthogonal Frequency Division Multiplexing (OFDM) technology for upstream and downstream communication. Figure 1a shows Mode 2 channelization of physical layer. The data transmission in the physical channel takes place on two-dimensional Resource Blocks (RBs), each of which consists of multiple subcarriers and a time slot. There exist two periodic time windows, namely the sensing window and the selection window (SW). In the sensing window, the vehicle senses the RBs in use by the other vehicles. In the SW, the vehicle selects or reserves appropriate RBs for data transmission. The subcarrier spacings of 30 kHz, 60 kHz, and 120 kHz are introduced to accommodate different communication requirements [6]. In the time domain, each channel occupies one time slot, whose duration varies depending on the subcarrier spacing. A preconfigured number of multiple continuous RBs within the same time slot form a sub-channel, which is used to transmit data and vehicle control information [1]. The data is organized in the Transport Block (TB) of the Physical Sidelink Shared Channel (PSSCH), and a TB can utilize one or multiple sub-channels [9].

Each TB has an accompanying Sidelink Control Information (SCI), which includes key information such as the Resource Reservation Interval (RRI), the size of the Reselection Counter (RC), number of RBs the TB occupies, the MCS of the TB, and the priority of the transmitted information [5]. A TB can only be

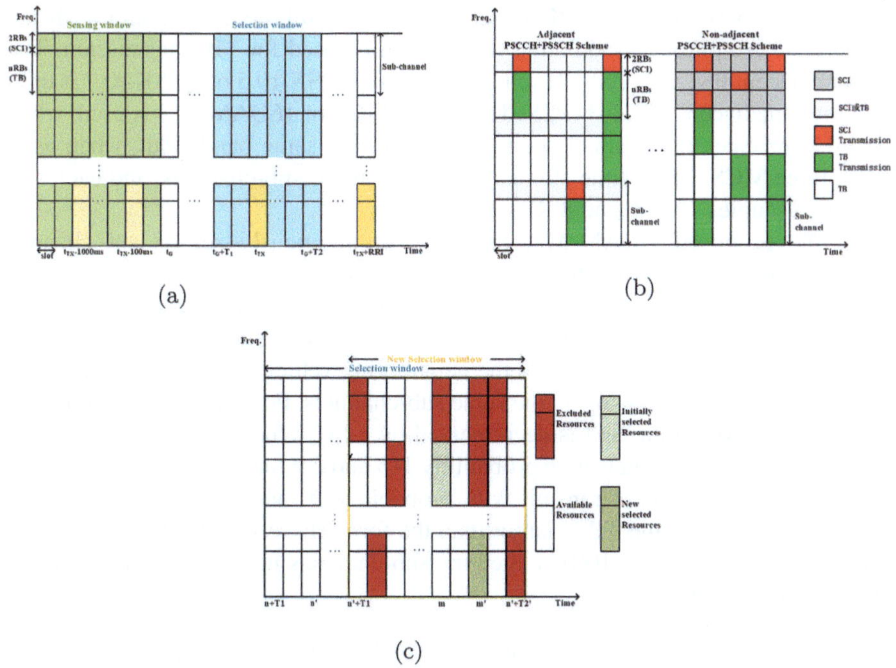

Fig. 1 Channelization, sub-channelization, and reevaluation mechanisms. (**a**) Channelization and resource allocation in Mode 2 ($T = 100\,\text{ms}$). (**b**) Schematic of the two sub-channelization. (**c**) Resource reevaluation mechanism

successfully decoded if the SCI is correctly received. The SCI occupies two RBs and is transmitted in the Physical Sidelink Control Channel (PSCCH). Each TB and its associated SCI are conveyed within the same time slot.

3 Proposed Adaptive Access Mechanism for Mode 2

The enhancement of the proposed ADAM-M2 lies into three aspects. It dynamically adjusts both sub-channelization schemes and MCSs for vehicles and enhances the existing SPS scheme. First, two sub-channelization schemes are defined as shown in Fig. 1b, namely adjacent PSCCH+PSSCH and nonadjacent PSCCH+PSSCH [10]. In the adjacent PSCCH+PSSCH scheme, SCIs and TBs are sent using neighboring RBs. The SCI takes up the first two RBs of the sub-channel, with the TB following. For the nonadjacent PSCCH+PSSCH scheme, the RBs are split into two resource pools. One pool is devoted to the transmission of SCIs, each of which occupies two RBs, and the second pool is used only for the transmission of TBs. In the existing Mode 2, only Quadrature Phase Shift Keying (QPSK) modulation is used. In our work, it is proposed that QPSK is used for SCI transmission, and both QPSK and Quadrature Amplitude Modulation (QAM) can be used for TB transmission. That is, for adjacent PSCCH+PSSCH scheme, MCS of QPSK-r0.5 or QPSK-r0.7 is used, and for nonadjacent PSCCH+PSSCH scheme, MCS of 16QAM-r0.5 is used, where r0.5/0.7 indicates the channel coding rate. With the same bandwidth, when using these three MCSs, each time slot contains 3, 4, and 6 sub-channels, and each sub-channel occupies 16, 12, and 8 RBs, respectively.

In the proposed ADAM-M2, the VUE autonomously selects sub-channels according to the improved SPS scheme and to set aside sub-channels for transmissions of packets, as shown in Fig. 1a, where the VUE selects sub-channels at time slot t_{TX} and reserves sub-channels at time slot $t_{TX} + RRI$ for its subsequent transmissions. It also broadcasts to the neighboring vehicles to prevent other VUEs from transmitting packets using the same sub-channels. For each TB transmitted by the VUE, the RC in the SCI is decremented by 1 [8]. When the RC reaches 0, the VUE reserves sub-channels with probability 1-p and continues to use the previously reserved sub-channels to transmit with probability p, where $p \in [0, 0.8]$. If the resource required by the TB is larger than the formerly reserved sub-channels or if the delay the TB reaches 100 ms, then the sub-channels need to be reselected. The steps are as follows:

Step 1: VUEs adjust communication parameters and data logging in different network coverage area. The base station broadcasts parameters to the VUEs in the coverage area, including quantity of sub-channels in the network, the size of the time-frequency resources for each sub-channel, the SPS period T, the probability of sub-channel reselection p, etc. When a VUE is outside the cellular coverage area and not able to receive the broadcast, it uses a collection of preset parameters. The VUE continuously scans to the sub-channels in the network and records the received signal strength.

Step 2: Delimit the SW and determine candidate resources. The SW is a time window containing the time slots within $t_G + T_1$ to $t_G + T_2$. If, within time slot t_G, the VUE needs to reserve a sub-channels for sending packets for periodic services, it performs resource selection within SW. t_G is the time slot in which the VUE intends to choose a new resource. T_1 is the duration required for the VUE to select candidate resources, $T_1 \leq 4$ time slots. T_2 denotes the delay time for the VUE to send the packet, to make $t_G + T_2$ satisfy the Packet Delay Budget (PDB) requirement for sending TB; the maximum delay should be controlled within 100 ms.

Step 3: Compose the candidate resource list L_1. The sensing window is utilized to filter out resources, within the SW, that may have been previously reserved by other VUEs, as well as those previously reserved by the VUE itself. The corresponding TB is found by the SCI received by the VUE, and the Reference Signal Received Power (RSRP) is measured on the TB. If the RSRP exceeds the set threshold, then it is concluded that this resource is already being used by other VUEs and is excluded. The number of resources L_1 is required to be greater than 20% of the number of all candidates in SW. Otherwise, the RSRP threshold is raised by 3 dB and repeatedly screening.

Step 4: Refine L_1 to get candidate resource list L_2. L_1 is refined by removing resources whose average Received Signal Strength Indicators (RSSIs) are highest till its resource number is equal to or as closely higher as possible than 20% of the resources number in the candidate list. The resulting list is named as L_2. The RSSI mean is evaluated over the $t_{TX} - T \cdot j$ time slots of the sensing window.

Step 5: Reevaluation mechanism to check the availability of initially selected resources. Choose an initially selected resource in time slot m from L_2, as shown in Fig. 1c. The VUE continues to sense transmissions from other VUEs at time slot m. Step 2 is performed again at time slot n' to delimit a new SW $\left[n' + T1, n' + T2'\right]$, where the availability of the initially selected resource is checked. If it is available, the packet is transmitted on this resource. Otherwise, steps 3 and 4 are performed to select a new resource for data transmission. Here, T_2' denotes the VUE packet transmission delay, $T_2 \min \leq T_2' \leq PDB - (n' - n)$, where n is the time slot in which the VUE delimits the SW when step 2 is performed last time, $T_{2\min}$ denotes the required minimum transmission delay, and $10 \leq T_{2\min} \leq 20$.

Step 6: Adaptive MCS selection. Define $CBR = \frac{N_E}{N}$, where N_E and N are the numbers of excluded resources and total channel resources in the SW, respectively. When $CBR \geq 0.8$, the MCS of QPSK-r0.5 is used, when $0.5 \leq CBR < 0.8$, the MCS of QPSK-r0.7 is used, and when $CBR < 0.5$, the MCS of 16QAM-r0.5 is used.

4 Analytical Models

To understand the PDR performance of VUEs in depth, four types of packet transmission errors are analyzed, namely, errors caused by half-duplex communication, errors caused by poor received sensing signal power, errors by poor received signal to noise ratio (SNR), and errors caused by packet collisions. A packet is considered to be successfully received if none of these transmission errors occurs. Define Φ_{HD}, $\Phi_{SEN}(d_{s,r})$, $\Phi_{SNR}(d_{s,r})$, and $\Phi_{COL}(d_{s,r})$ as the error probabilities of four errors, respectively, where $d_{s,r}$ is the distance between sender and receiver. Then $PDR(d_{s,r})$ is

$$\begin{aligned} PDR(d_{s,r}) = (1 - \Phi_{HD}) \cdot \left(1 - \Phi_{SEN}(d_{s,r})\right) \\ \cdot \left(1 - \Phi_{SNR}(d_{s,r})\right) \cdot \left(1 - \Phi_{COL}(d_{s,r})\right) \end{aligned} \tag{1}$$

4.1 Error Probability Caused by Half-duplex

The half-duplex transmission occurs when both vehicles transmit in the same time slot. The error probability depends on the packet transmission rate per vehicle λ and the number of time slots per second Ts, which is related to subcarrier spacing. It can be expressed as

$$Ts = 1000 \cdot 2^n, \quad 0 \leq n \leq 2 \tag{2}$$

where $n = 0, 1$, and 2 presents the case that the subcarrier spacing is 15, 30, and 60 kHz, respectively. Thus, Φ_{HD} can be derived as

$$\Phi_{HD} = \frac{\lambda}{Ts} = \frac{\lambda}{1000 \cdot 2^n}, \quad 0 \leq n \leq 2 \tag{3}$$

4.2 Error Probability Caused by Poor Received Signal

This error happens when the received power is under the sensing power threshold. It is therefore related to the transmit power P_t, the sensing power threshold P_{SEN}, the distance from the sender vehicle to the receive vehicle $d_{s,r}$, and the signal attenuation over the radio path. The power of the received signal $P_r(d_{s,r})$ in dB at distance $d_{s,r}$ is

$$P_r(d_{s,r}) = P_t - PL(d_{s,r}) - SH \tag{4}$$

in which $PL(d_{s,r})$ represents the path loss at distance $d_{s,r}$ in dB. SH is the shadowing effect that in dB is modeled as a normal distribution with standard deviation of σ and zero mean. This error probability $\Phi_{SEN}(d_{s,r})$ therefore can be expressed as

$$\Phi_{SEN}(d_{s,r}) = \int_{-\infty}^{P_{SEN}} f_{p_r, d_{s,r}}(p)dp \tag{5}$$

For which $f_{p_r, d_{s,r}}(p)$ represents the probability density function of the received power at $d_{s,r}$, denoted by

$$f_{p_r, d_{s,r}}(p) = \frac{1}{\sigma\sqrt{2\pi}} \exp\left(-\left(\frac{P_t - PL(d_{s,r}) - p}{\sigma\sqrt{2}}\right)^2\right) \tag{6}$$

Hence,

$$\Phi_{SEN}(d_{s,r}) = \frac{1}{2}\left(1 - erf\left(\frac{P_t - PL(d_{s,r}) - P_{SEN}}{\sigma\sqrt{2}}\right)\right) \tag{7}$$

in which erf is the error function. Given a distance of d, the Packet Sensing Ratio (PSR) may be defined as $1 - \Phi_{SEN}(d)$ and expressed as

$$PSR(d) = \frac{1}{2}\left(1 + erf\left(\frac{P_t - PL(d) - P_{SEN}}{\sigma\sqrt{2}}\right)\right) \tag{8}$$

4.3 Error Probability Caused by Poor Received SNR

Even though the received power exceeds the sensing threshold, SNR may still be insufficient for correct packet reception. This error depends on transmit power, SNR threshold, distance between vehicles, link propagation, and the MCS used. The $SNR(d_{s,r})$ in dB can be denoted by

$$SNR(d_{s,r}) = P_r(d_{s,r}) - N_0 = P_t - PL(d_{s,r}) - SH - N_0 \tag{9}$$

In this case N_0 is the background noise power. $\Phi_{SNR}(d_{s,r})$ therefore can be defined as

$$\Phi_{SNR}(d_{s,r}) = \sum_{s=-\infty}^{+\infty} BL(s) \cdot f_{SNR|P_r>P_{SEN},d_{s,r}}(s) \tag{10}$$

in which $f_{SNR|P_r>P_{SEN},d_{s,r}}(s)$ denotes the probability density function of the SNR at $d_{s,r}$ given that $P_r > P_{SEN}$, and $BL(s)$ denotes the Block Error Rate (BLER) in the lookup table for SNR of s for a specific MCS.

$$f_{SNR|P_r>P_{SEN},d_{s,r}}(s) = \begin{cases} \frac{f_{SNR,d_{s,r}}(s)}{1-\Phi_{SEN}} & if\ P_r > P_{SEN} \\ 0 & if\ P_r \leq P_{SEN} \end{cases} \tag{11}$$

4.4 Error Probability Caused by Packet Collision

This error is related to the transmission parameters, the resource allocation scheme, the distance between the sender and receiver, and the traffic density. When two vehicles are transmitting on the common sub-channel, the interference produced causes inadequate Signal-to-Interference and Noise Ratios (SINRs) leading to the error. Therefore, $\Phi_{COL}(d_{s,r})$ can be written as

$$\Phi_{COL}(d_{s,r}) = 1 - \prod_i (1 - \Phi^i{}_{COL}(d_{s,r}, d_{s,i}, d_{i,r})) \tag{12}$$

$$\Phi^i{}_{COL}(d_{s,r}, d_{s,i}, d_{i,r}) = P_{SIM}(d_{s,i}) \cdot P_{INT}(d_{s,r}, d_{i,r}) \tag{13}$$

where $\Phi^i{}_{COL}(.)$ denotes the probability of packet collision caused by interfering vehicle v_i, $P_{SIM}(d_{s,i})$ denotes the probability that the sender vehicle v_s and v_i use the identical resources for transmission, and $P_{INT}(d_{s,r}, d_{i,r})$ denotes the probability that the interference generated by v_i is exceeding a threshold. Then the $SINR$ at the receive vehicle v_r is

$$SINR(d_{s,r}, d_{i,r}) = P_r(d_{s,r}) - P_i(d_{i,r}) - N_0 \tag{14}$$

In which $P_i(d_{i,r})$ is the power accepted by v_r from v_i. So the probability of packet reception failure $P_{SINR}(d_{s,r}, d_{i,r})$ equals

$$P_{SINR}(d_{s,r}, d_{i,r}) = \sum_{s=-\infty}^{+\infty} BL(s) \cdot f_{SINR|P_r > P_{SEN}, d_{s,r}, d_{i,r}}(s) \tag{15}$$

Since the probability has already been considered, normalization is required as follows:

$$P_{INT}(d_{s,r}, d_{i,r}) = \frac{P_{SINR}(d_{s,r}, d_{i,r}) - \Phi_{SNR}(d_{s,r})}{1 - \Phi_{SNR}(d_{s,r})} \tag{16}$$

So as to calculate $P_{SIM}(d_{s,i})$, we must consider the resource distributions of v_s and v_i. We define these variables below. N_E is the number of total excluding resources in SW. N_A is the quantity of total assignable resources in SW. N_c represents the count of total candidate resources in SW. $C_E(.)$ is the total of commonly excluded resources for v_s and v_i in SW. $C_A(.)$ is the amount of commonly available resources for v_s and v_i in SW. $C_C(.)$ is the quantity of common candidate resources for v_s and v_i in SW. $P_s(.)$ is the probability that v_s and v_i do not consider the other's transmission before selecting a new resource. S_{PSR} represents the mean number of vehicles detected by a single vehicle in the SW, assuming no packet collisions occur. $R_{PSR}(d_{s,i})$ refers to the autocorrelation of the PSR function at $d_{s,i}$.

$P_{SIM}(d_{s,i})$ depends on $C_C(d_{s,i})$, N_C and the probability $P_s(d_{s,i})$. Therefore,

$$P_{SIM}(d_{s,i}) = P_s(d_{s,i}) \cdot \frac{C_C(d_{s,i})}{N_C^2} \tag{17}$$

$$P_s(d_{s,i}) = 1 - \left(1 - \frac{1}{\tau}\right) \cdot PSR(d_{s,i}) \tag{18}$$

where τ depends on λ, if v_s and v_i are apart from the sensing range of each other, $PSR\left(d_{s,i}\right) = 0$, otherwise, $PSR\left(d_{s,i}\right) = 1$. $C_A(d_{s,i})$ can be expressed as a function of the total resource N, N_E, and $C_E(d_{s,i})$. Therefore, the following can be derived:

$$C_A(d_{s,i}) = N - 2 \cdot N_E + C_E(d_{s,i}) \tag{19}$$

$$C_C(d_{s,i}) = C_A(d_{s,i}) \cdot \left(\frac{N_C}{N_A}\right)^2 \tag{20}$$

Depending on the MCS used, the amount of sub-channels in each time slot, S, can be 3, 4, 6. Therefore the overall amount of resources N in SW is

$$N = Ts \cdot \frac{S}{\lambda} = 1000 \cdot 2^n \cdot \frac{S}{\lambda}, \quad 0 \leq n \leq 2 \tag{21}$$

$$N_C = 0.2 \cdot N \tag{22}$$

In order to calculate $C_A(d_{s,i})$, it is necessary to figure out $C_E(d_{s,i})$ and N_E, which is determined by the transmit power, N, traffic density β in vehicle/m, etc.

$$N_E = \frac{S_{PSR}}{2} + \sum_{k=1}^{S_{PSR/2}} \max\left(1 - \frac{k}{N - S_{PSR/2}}, 0\right) \tag{23}$$

$$S_{PSR} = \beta \cdot \sum_{i=-\infty}^{+\infty} PSR(i) \tag{24}$$

$$C_E(d_{s,i}) = \frac{R_{PSR}(d_{s,i})}{R_0}\left(\frac{\beta \cdot N_E \cdot R_0}{S_{PSR}} - \frac{N_E^2}{N}\right) + \frac{N_E^2}{N} \tag{25}$$

$$R_{PSR}(d_{s,i}) = \sum_{j=-\infty}^{+\infty} PSR\left(\frac{j}{\beta} + d_{s,i}\right) \cdot PSR\left(\frac{j}{\beta}\right) \tag{26}$$

$$R_0 = R_{PSR}(0) \tag{27}$$

$P_{SIM}(d_{s,i})$ can be calculated by considering Eqs. (17)–(27).

5 Model Validation and Simulation

The simulated system scenario is based on the 3GPP Manhattan urban scenario and implemented in MATLAB. The vehicle locations are deployed on the road according to the spatial Poisson distribution process, and the vehicles travel at a constant speed within the lanes. The key simulation parameters and their settings are shown in Table 1.

Table 1 Simulation parameters and settings

Parameters	Settings
Packet size	190 bytes
Subcarrier spacing	30 kHz
Channel bandwidth	20 MHz
Packet tx frequency (λ)	10,20 Hz
Transmission power (P_t)	20,23 dBm
Number of sub-channels per time slot	3,4,6

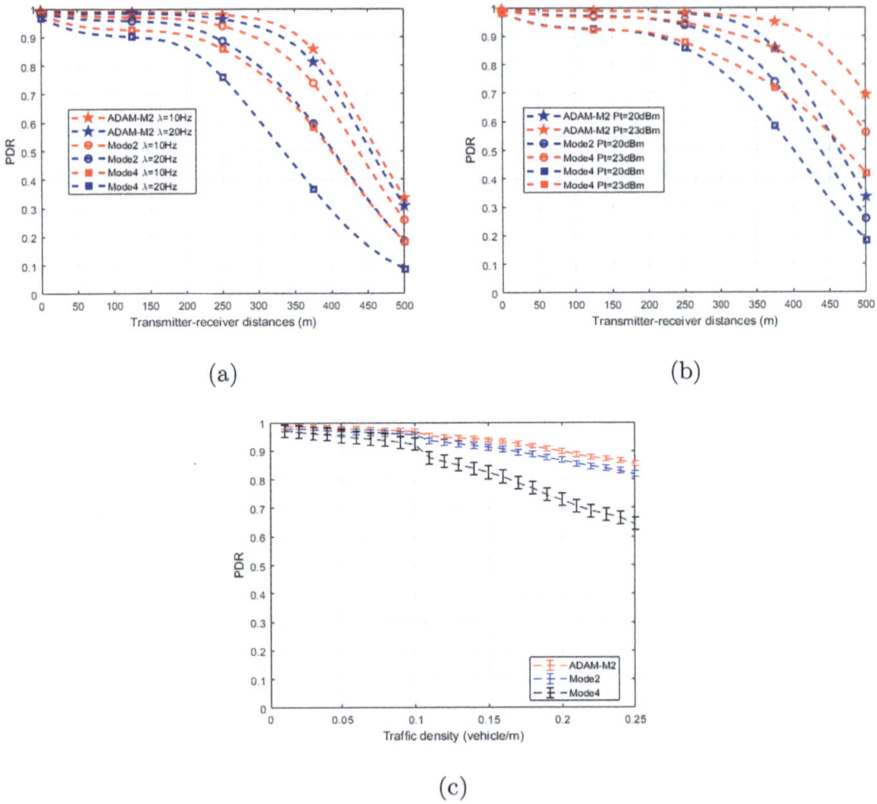

Fig. 2 Comparison of the three access mechanisms. (**a**) Comparison plot of the three access mechanisms with different λ. (**b**) Comparison plot of the three access mechanisms with different P_t. (**c**) Comparison plot of the three access mechanisms at different traffic densities

Figure 2a shows the PDR performance versus the sender-receiver distance for the three access mechanisms with different packet transmission frequencies. As expected, the PDR performance is better when each vehicle transmission is less frequent because of less channel load and less disruptions faced by vehicles. It is also evident that the proposed ADAM-M2 shows better performance regardless of the packet transmission frequencies.

Figure 2b show the PDR performance versus the sender-receiver distance with different vehicle transmission powers. It can be seen that an increase in P_t helps improve the PDR, especially over longer communication distances. Moreover, it is evident that the ADAM-M2 consistently outperforms Mode 4 and Mode 2 under different transmit powers. However, it should be noted that a higher P_t leads to greater energy consumption. Therefore, further study is worth carrying out to jointly optimize the package receiving performance and power consumption.

Figure 2c compares the performance of ADAM-M2 with Mode 2 and Mode 4 under different traffic densities. It shows that ADAM-M2 serves more VUEs at the same PDR, effectively reducing resource congestion. The error bars indicate that ADAM-M2 has the smallest performance fluctuation, making it the most stable.

6 Conclusion

In this chapter, we introduce for the first time an ADAM-M2 scheme as an enhancement to the existing 3GPP Mode 2 for information access and transmission in C-V2X systems. The scheme allows VUEs to dynamically select the most appropriate access mechanism based on real-time traffic density. The channel busy rate is utilized as the criterion to guide vehicles to adaptively choose the most appropriate resources. In the resource allocation phase, different sub-channelization schemes are chosen for different MCSs according to the traffic density. Simulation results show that the proposed ADAM-M2 can effectively improve network performance and optimize resource utilization under different traffic conditions, which can help realize intelligent transport systems. In the future, this can be combined with Ultra-Reliable Low-Latency Communication (URLLC) technology in 6G networks to further improve communication performance in high-density environments and ensure that adaptive tuning can be realized with very low latency.

References

1. Saad, M.M., Tariq, M.A., Siddiqa, A., Kim, D.: Analytical models for distributed resource allocation in nr-v2x mode 2. IEEE Int. Things J., 1–1 (2023). http://dx.doi.org/10.1109/JIOT.2023.3328885.
2. Wang, P., Di, B, Zhang, H., Bian, K., Song, L.: Cellular V2X communications in unlicensed spectrum: Harmonious coexistence with VANET in 5G systems. IEEE Trans. Wirel. Commun. **17**(8), 5212–5224 (2018)
3. Guo, C., Li, D., Chen, X., Zhang, G.: An adaptive V2R communication strategy based on data delivery delay estimation in VANETs. Veh. Commun. **34**(C), 100444 (2022)
4. Charpentier, V., Slamnik-Krijestorac, N., Marquez-Barja, J., Costa, C.: A proposal on a connected automated mobility (cam) communication system for (U)AVs. In: Proceedings of the 2022 ACM Conference on Information Technology for Social Good, GoodIT '22, pp. 225–230, New York, NY (2022)

5. Gonzalez-Martín, M., Sepulcre, M., Molina-Masegosa, R., Gozalvez, J.: Analytical models of the performance of c-v2x mode 4 vehicular communications. IEEE Trans. Vehic. Technol. **68**(2), 1155–1166 (2019)
6. Garcia, M.H.C., Molina-Galan, A., Boban, M., Gozalvez, J., Coll-Perales, B., Şahin, T., Kousaridas, A.: A tutorial on 5G NR V2X communications. IEEE Commun. Surv. Tutor. **23**(3), 1972–2026 (2021)
7. Nabil, A., Kaur, K., Dietrich, C., Marojevic, V.: Performance analysis of sensing-based semi-persistent scheduling in c-v2x networks. In: 2018 IEEE 88th Vehicular Technology Conference (VTC-Fall), pp. 1–5 (2018)
8. Todisco, V., Bartoletti, S., Campolo, C., Molinaro, A., Berthet, A.O., Bazzi, A.: Performance analysis of sidelink 5G-V2X mode 2 through an open-source simulator. IEEE Access **9**, 145648–145661 (2021)
9. Bagheri, H., Noor-A-Rahim, M., Liu, Z., Lee, H., Pesch, D., Moessner, K., Xiao, P.: 5g nr-v2x: Toward connected and cooperative autonomous driving. IEEE Commun. Stand. Mag. **5**(1), 48–54 (2021)
10. Molina-Masegosa, R., Gozalvez, J.: LTE-V for sidelink 5G V2X vehicular communications: a new 5G technology for short-range vehicle-to-everything communications. IEEE Vehic. Technol. Mag. **12**(4), 30–39 (2017)

User Classification Model Based on NSGA-II Optimized LightGBM

Peiwen Yu, Yang Zhao, Xuejian Wang, Ruikang Wang, and Tian Bai

1 Introduction

The banking industry is the backbone of China's financial sector, accounting for over 90% of total financial assets and maintaining the closest ties with the real economy. To comprehensively promote the development of the real economy, effective measures are required to continuously enhance the financial sector's ability to serve the real economy.

Jilin Bank is an important urban commercial bank in the Northeast region of China. It has consistently innovated its existing credit card marketing strategies, particularly focusing on credit card installment services. Credit card installment services can cover various medium-to large-scale consumption scenarios and serving the profit objectives of financial institutions. Credit card outbound calling is a significant method for promoting credit card installment services. Traditional credit card outbound calling often leads to low outbound calling efficiency and a high failure rate. Bank customer data possesses characteristics such as globality, multidimensionality, and correlations, making the utilization of machine learning methods for data analysis a hot topic.

P. Yu
College of Software, Jilin University, Changchun, China

Y. Zhao · X. Wang
Financial Technology Department, BANK OF JILIN CO., LTD, Changchun, China

R. Wang · T. Bai (✉)
College of Computer Science and Technology, Jilin University, Changchun, China
e-mail: baitian@jlu.edu.cn

X. Jia et al. (eds.), *2nd International Conference on Big Data, IoT, and Cloud Computing*, EAI/Springer Innovations in Communication and Computing,
https://doi.org/10.1007/978-3-031-93570-1_20

LightGBM is a gradient boosting framework built on top of GBDT [1]. It achieves efficient training on large-scale datasets by utilizing techniques such as histogram-based growth, leaf-wise growth, gradient-based learning, and multi-threading support [2]. Given the large size of the dataset used in this study and limitations in the production environment and hardware, using the LightGBM model can establish a more accurate classification model under these conditions.

In the dataset used in this study, user behavior exhibits high dispersion and high dimensionality. Traditional linear feature reduction and clustering algorithms based on Euclidean distance are not suitable for high dimensional discrete data. The high dimensionality of data not only increases computational and storage costs but also introduces noise and issues related to insufficient data samples, significantly impacting the performance of traditional algorithms [3]. For high dimensional discrete data, the NSGA-II algorithm is a metaheuristic population-based intelligent algorithm. Currently, these adaptive search algorithms are widely applied in feature selection [4]. The algorithm's processing is based on a mathematical simulation of biological genetic evolution rather than computing the information of the data features themselves. To apply it to practical problems, it is necessary to specify appropriate iteration counts and fitness function [5]. This paper begins by preprocessing and performing preliminary data handling. Next, it employs the NSGA-II algorithm for feature selection, identifying the feature set to be used in the model. Subsequently, the LightGBM algorithm is applied to train on the final feature set using a dataset consisting of millions of customers. The model predicts credit card customer potential scores. A scoring threshold is manually specified, and customers are categorized based on this threshold. Customers with high potential scores are selected for a priority outbound call list, which is then provided to a remote outbound calling team.

2 Designing and Constructing the Dataset

The data for this research is a comprehensive dataset provided by Jilin Bank Credit Card Center, which includes a customer relationship table (CRM), monthly customer billing table (BIL), and monthly installment table (INST), each spanning 5 months of data, from 2023/2 to 2023/6. The data volume of each table has reached one hundred thousand levels.

Because in the task of this paper, it is necessary to predict the installment situation, so when constructing the dataset, it is necessary to associate the monthly billing table and customer relationship table of the previous month with the monthly installment table of the next month to generate training labels.

3 Model Design and Construction

3.1 *Model Design and Construction*

By analyzing and summarizing the content of the BIL table, CRM table, and INST table, the data in these three tables are defined as scalar data. This portion of the data is suitable for processing using machine learning binary classification prediction algorithms. Given the data's relatively discrete and high dimensional heterogeneous characteristics, the LightGBM algorithm is well-suited for handling it.

To further reduce algorithmic overhead, data preprocessing is required before entering the model. In addition to common data cleaning and feature engineering used in general scenarios, dimensionality reduction is necessary. In this paper, an improved NSGA-II algorithm is used with LightGBM as the algorithm iterator to perform feature selection on the data. The filtered new feature set is then input into the LightGBM model for training, yielding the final results. The overall model proposed in this paper will be referred to as the "L model" in the following sections.

3.2 *Model Construction*

Data Preprocessing

First, preliminary processing of data is performed for each individual table. The string-type fields and non-string-type fields are handled separately. For the remaining object-type fields, which were found to be mixed types containing both strings and empty strings upon inspection, all empty strings are converted to null values for uniform processing.

Regarding other int/float-type fields, visual analysis is conducted using the matplotlib library to observe data distributions for subsequent processing. Missing values are filled using mean imputation, and outliers are replaced using upper and lower bounds.

The *describe()* function is employed to analyze the variance and mean of each column in each table, while the value counts attribute of the dataframe is used to observe the number of missing values in each column. Combining these results with previous visualizations, columns with missing value proportions exceeding 99.9% and variance less than 0.001 are deleted, further optimizing data quality.

Prior Processing

For non-natural language string fields, they are currently processed using ordinal encoding. This approach serves two purposes: firstly, it allows for the clear differentiation of various categories represented by string fields, particularly for those indicating item or order numbers and other strings that have a significant association with other fields. Secondly, ordinal encoding preserves some of the relational information between fields due to its encoding characteristics.

In this context, the following prior features are processed: credit card types, age, assets, and addresses. They are binarized and then numerically encoded after setting up range bins.

Feature Selection

After the preliminary screening and prior processing, the dimensionality of the data remains high, which can lead to high computational costs in subsequent modeling and potentially result in underfitting.

The entire dataset is divided into two parts: one part is the k dimensional feature dataset M, where the feature set C has a cardinality of $|C| = k$. The other part is a randomly sampled scoring dataset P, where $P < < M$. In this study, the NSGA-II algorithm is used for one round of feature selection.

The number of evolutionary generations is set to T, and the elite retention rate is set to a. For $t \in [0, T]$, the initial generation $t = 0$ is set. Initialize the population R_0 with n individuals in the generation_0, where each individual X_{0i} is represented, and $i \in [1, n]$. From the feature set C, randomly select $[1, |C|]$ features to compose X_{0i}, where $\{X_{01}, X_{02}, ..., X_{0i}, ..., X_{0n} \mid X_{0i} \subseteq C, i \in [1, n]\}$. Using the scoring dataset P, generate individual scoring sets P_{ti} with features X_{ti} and split them into training sets P_{ti_train} and validation sets P_{ti_verify}. Train P_{ti_train} using LightGBM and use P_{ti_verify} for prediction, generating prediction results. Use the AUC (Area Under the Curve) as the fitness function. Evaluate the prediction results based on the labels in P_{ti_verify} and calculate the AUC as the fitness D_{ti} of the individual $X_{ti..}$

The following describes the genetic evolution of each generation in the population. For each individual, the objective function is used to evaluate the quality of the feature subset, resulting in a fitness value. Nondominated sorting is then performed on the population, categorizing individuals into different ranks and establishing the Pareto front among individuals based on dominance relationships, aiming to find feature subsets as shown in formulas 31 and 32:

$$f_j(C_{t0}) >= f_j(C_{t1}) \quad j = 1, 2, \ldots, m; \tag{1}$$

$$f_j(C_{t0}) > f_j(C_{t1}) \quad j = 1, 2, \ldots, m; \tag{2}$$

C_{t0} is said to dominate C_{t1}.To find the Pareto front, solutions with the same dominance relationship are identified. Calculating the crowding distance between individuals is used to maintain diversity. Nondominated sorting and crowding distance information are employed for selection, preserving a portion of superior individuals as parents for crossover and mutation. The parent and offspring individuals are then combined, keeping the population size constant, completing the population update. The process of genetic evolution is repeated to generate a new population. Iteration stops when the fitness change is small or the set number of iterations is reached. The filtered feature subset is returned as the result.

3.3 *Building a LightGBM Model*

After processing the dataset as described earlier, it is split into training and testing datasets. Parameter settings are performed for LightGBM, which has many adjustable hyperparameters. To create a LightGBM model, first create a LightGBM model object, pass the dataset and hyperparameters to the model, and then call the model's training function, usually the *train()* method, and provide it with the training dataset and corresponding parameters. After a series of parameter adjustment experiments, the main parameter values set in this paper are as follows: learning_rate: 0.03, num_leaves: 31, max_depth: 1, min_data_in_leaf: 20, max_bin: 8191, and the number of iterations: 800.

4 Experimental Results and Analysis

4.1 *Introduction to the Experimental Dataset*

We conducted a series of experiments on both public datasets and a real banking dataset. For the Jilin Bank's real dataset, we split the data from the April learning cycle into training and testing sets. We used the data from the May learning cycle as a separate prediction set for a series of experiments. Regarding the public dataset, it was provided by researchers at Imperial College London. Each feature in this dataset, like the Jilin Bank dataset, exhibits high discreteness and diverse data types. Each data point is independent of the others. During the experiments, we classified any loss greater than 0 as 1 to transform it into a binary classification dataset suitable for this problem.

4.2 *Ablation Experiment*

In order to test the performance improvement of the NSGA-II algorithm and sparse clustering algorithm on the model, conducted ablation experiments. The results are shown in Table 1.

From the data in the tables, it can be observed that after using the NSGA-II algorithm for feature selection, the model's accuracy slightly improved on the bank

Table 1 Sensitivity experiment on real bank dataset

Model	Average Runtime(ms)	Total Runtime(ms)	F1	Recall	Accuracy	Auc_roc
LightGBM	399	173,426	0.753459	0.675873	0.971240	0.998210
Lmodel	**201**	**78,284**	**0.585268**	**0.601151**	**0.970412**	**0.998959**

Table 2 Comparative experiment of GBDT on public datasets

Model	Dataset	F1	Recall	Accuracy
Catboost	Test set	0.051458	0.0414973	0.901143
Adaboost	Test set	0.052619	0.047807	0.902282
L model	**Test set**	**0.091800**	**0.051941**	**0.903389**

Table 3 Comparative experiment of GBDT on real bank dataset

Model	Dataset	F1	Recall	Accuracy
Catboost	Test set	0.151414	0.221142	0.963387
	Prediction set	0.137331	0.075279	0.960298
Adaboost	Test set	0.372001	0.316139	0.963358
	Prediction set	0.365761	0.294506	0.959138
L model	**Test set**	**0.638872**	**0.657893**	**0.970898**
	Prediction set	**0.585268**	**0.601151**	**0.970412**

dataset, but it decreased on the public dataset. However, the running efficiency improved significantly in both cases.

4.3 Comparative Experiments

In order to evaluate the training performance of our model, we compared it with several state-of-the-art binary classification algorithms, including Xgboost, Adaboost, NGBoost, and Catboost. The results are shown in the table below (Tables 2 and 3):

It can be seen from the two tables above. If feature selection successfully identifies a subset of features that better describe the data, it helps to reduce noise and redundant features, thus enabling more focused learning of these complex relationships. Therefore, in the public dataset, due to the lack of feature selection using the NSGA-II algorithm, the dataset has more noise interference and redundant features, resulting in lower performance of other algorithms.

4.4 Real-World Application

We deployed the algorithm to the outbound platform of the bank's credit card center, which took 3 months from the initial version to a mature model. Each month, from

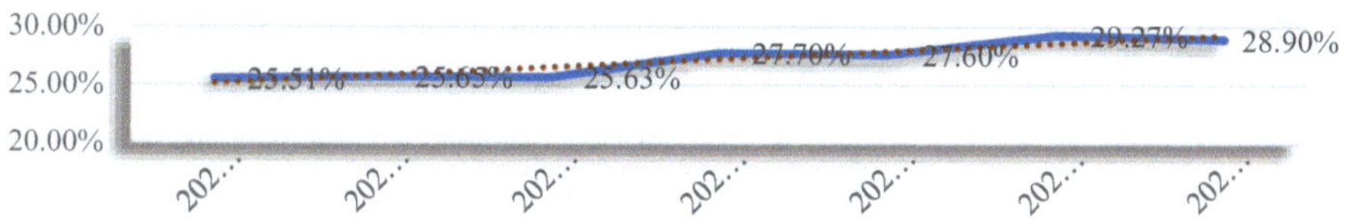

Fig. 1 Installment application rate from March to June

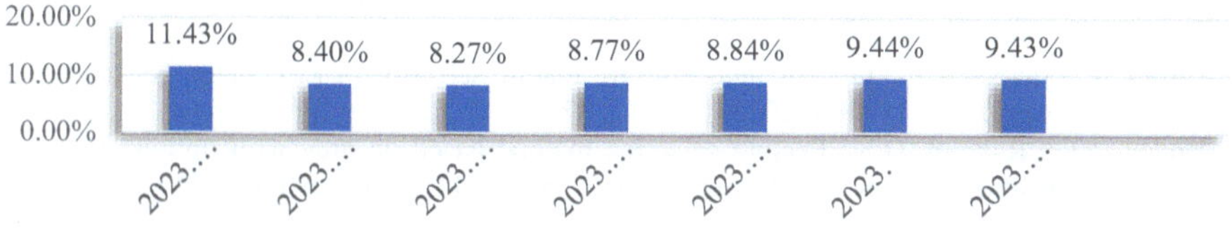

Fig. 2 Outbound installment success rate

the first to the 31st (30th), is referred to as a natural month, and from the 15th of 1 month to the 15th of the next month is referred to as a billing cycle (Figs. 1 and 2).

Since the initial deployment of the outbound model in March 2023, the installment application rate has gradually increased. This increase was particularly noticeable in the first month of model deployment, where the application rate increased by 2.07%. Furthermore, when comparing the natural months of May and March, there was a 3.76% increase in installment application rates. Similarly, when comparing billing cycles in May with those in March, there was a 3.27% increase. In comparing the success rate between May and March natural months, there was a 1.04% increase, and when comparing the billing cycles of May and March during the same period, there was a 1.16% increase in success rates.

5 Conclusion

This paper provides a viable outbound effectiveness improvement model for urban commercial banks. It explores the combination of the NSGA-II algorithm and the LightGBM algorithm to address the comprehensive problem of feature dimensionality reduction and binary classification on high dimensional datasets. Experimental results demonstrate that the proposed approach performs well on multiple datasets and outperforms traditional feature selection and classification methods in terms of performance.

References

1. Qiang, Y., Yan, F., Li, G., Wenyu, W., Sichen, W., Qiangyu, L.: XBand radar attenuation correction method based on LightGBM algorithm. Remote Sensing. **15**(3), 864 (2023)
2. Ying, L., Haoran, Z., Jieguang, S., Yahui, T.: Digital inclusive finance and family wealth: evidence FROM LightGBM approach. Sustainability. **14**(22), 15363 (2022)
3. Azam, D., Mahboubeh, S., Golnoush, A., Arash, K.: Empirical analyses of genetic algorithm and grey wolf optimiser to improve their efficiency with a new multiobjective weighted fitness function for feature selection in machine learning classification: the roadmap. J. Exp. Theor. Artific. Intell. **35**(2), 171 (2023)
4. Aurora, S., Francesca, G.: Genetic algorithm for feature selection of EEG heterogeneous data. Expert Syst. Appl. **217**, 119488 (2023)
5. Pandey, A.C., Rajpoot, D.S.: Feature selection method based on grey wolf optimization and simulated annealing. Recent Adv. Comput. Sci. Commun. **14**(2), 635–646 (2021)

Efficient Broadcast Blind Multi-Signature Scheme Based on RSA

Feng Zhang, Yiwei Lin, Zhaochuan Li, and Yangong Wang

1 Introduction

With the acceleration of digital transformation and the widespread application of new technologies, data exchange is becoming increasingly frequent, and the application of digital signatures is also becoming more common. Digital signatures utilize private key encryption to authenticate messages, and according to the different intended uses, digital signatures with special properties have been developed, such as blind signatures, proxy signatures, and multi-signatures.

Chaum [1] first introduced the blind signature based on RSA. A blind signature permits a signer to sign a message without knowing its content and ensures that the signer cannot establish a connection between the signed message and the signing process. With this unlinkability and untraceability, the blind signature is widely used in many e-commerce protocols [2–4] and voting protocols [5, 6]. For more related papers, one can refer to [7, 8], and the references therein. The multi-signature, proposed by Itakura and Nakamura [9], is another special digital signature that allows multiple signers to collectively sign a message using a single compact signature. Interested in its salient feature, diverse literature has been published on related fields. Harn and Kresler [10], Bellare and Neven [11], Bagherzandi and Jarecki [12], Tsai et al. [13], and Hohenberger and Waters [14] proposed multi-signature schemes based on discrete logarithm problem, and Boneh et al. [15], Chen [16], Ristenpart and Yilek [17], He et al. [18], and Boneh et al. [19] proposed multi-signature schemes based on elliptic curve cryptography and bilinear paring. Multi-signatures can also be classified into sequential multi-signature schemes and broadcast multi-signature schemes based on the different signature processes.

F. Zhang · Y. Lin (✉) · Z. Li · Y. Wang
Inspur Software Technology Co. Ltd., Jinan, China
e-mail: linyiwei@inspur.com

X. Jia et al. (eds.), *2nd International Conference on Big Data, IoT, and Cloud Computing*, EAI/Springer Innovations in Communication and Computing,
https://doi.org/10.1007/978-3-031-93570-1_21

From the practical applications of e-commerce, in many cases, not only blind signatures but also multi-signatures are needed. For example, in a digital currency system, the electronic cash may be valid after several departments of the bank sign it blind at the same time. Another example is the electronic bidding process, and in order to protect the interests of bidders and ensure that the content of bidding documents will not be leaked before the opening of bidding, it is necessary for bidding units, procurement agencies, supervision departments, and other parties to sign bidding documents blind. In these cases, multi-blind signatures play an important role.

In 1995, Petersen et al. [20] introduced a blind multi-signature scheme based on discrete logarithms. It combines characteristics of multi-signatures and blind signatures. When multiple signers cooperate to sign the same message, they cannot see the specific content of this message, which can protect rights and interests of the signature requester. And many blind multi-signature schemes based on elliptic curves are studied, such as [21, 22].

In this chapter, considering the widespread implementation and application of RSA, we propose an efficient broadcast blind multi-signature scheme based on RSA. This scheme does not need any secure channel, and the final identity private keys are calculated by Key Generation Center (KGC) and signers, so that this scheme can not only realize the implicit verification of signers' private keys but also resist the attack of malicious KGC. In Sect. 2, we introduce our new scheme, and then the security of our scheme is proved under the RSA one-way problem in the random oracle model in Sect. 3. Finally, the conclusion is given in Sect. 4.

2 A New Scheme

In this section, we introduce an efficient broadcast blind multi-signature. In order to strengthen the privacy protection of the signature requester and reduce the risk of leakage of the message to be signed, a blind step is added, and the blinded message is then sent to multiple signers for digital signature.

Our scheme includes five stages: (1) initialization, (2) key generation, (3) requesting, (4) signing, and (5) extraction and verification. In the initialization stage, the KGC disseminates some essential information. In the key generation stage, KGC generates verifiable private keys for each signer based on their respective random secret values and identity information. All these random secret values of signers are aggregated to form part of the final signature. In the requesting stage, the requester prepares the blinded message using some random blind factor and sends it to KGC. KGC signs this blinded message using private key, and these signers, then, sign this data with aggregated secret values and their identity private keys. Finally, the requester collects all signed data from signers and derives the final signature. For the convenience of readers, we illustrate the above stages in Fig. 1.

Next, we introduce our new scheme step by step.

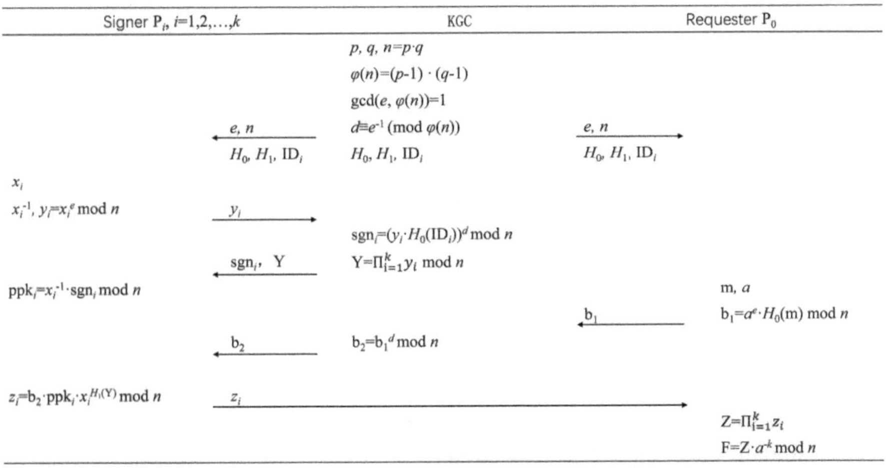

Fig. 1 The signature process of our proposed scheme

(1) **Initialization**

- KGC randomly chooses two large primes p and q and computes $n = p \cdot q$ and $\varphi(n) = (p-1)(q-1)$.
- KGC randomly chooses e, such that $1 < e < \varphi(n)$ and $\gcd(e, n) = 1$, and then calculates d satisfying $d = e^{-1} \mod \varphi(n)$. KGC publishes public key (e, n) and keeps (d, p, q) secretly as private key.
- KGC publishes three strong collision-resistant hash functions (such as SHA-256 [23]) $H_0 : \{0, 1\}^* \rightarrow \mathbb{Z}_n^*$, $H_1 : \{0, 1\}^* \rightarrow \mathbb{Z}_n^*$, $H_2 : \{0, 1\}^* \rightarrow \{0, 1\}^l$, $l \leq \log_2 n$.
- For all k signers P_i, $i = 1, \ldots, k$, KGC also publishes their identity information ID_i.

(2) **Key generation**

- For each $i = 1, \ldots, k$, signer P_i randomly selects a number $x_i \in \mathbb{Z}_n^*$ as his/her secret value, calculates x_i^{-1} and $y_i = x_i^e \mod n$, and then sends y_i to KGC.
- KGC computes $sgn_i = (y_i \cdot H_0(ID_i))^d \mod n$, $Y = \Pi_{i=1}^k y_i \mod n$, and it sends the tuple (sgn_i, Y) to the signer P_i, $i = 1, \ldots, k$.
- After receiving the tuple (sgn_i, Y), each signer P_i computes his/her own identity private key $ppk_i = x_i^{-1} \cdot sgn_i \mod n$.

(3) **Requesting**

- The requester P_0 randomly chooses a blind factor a, where $a \in \mathbb{Z}_n^*$. For message m, P_0 then computes $b_1 = a^e \cdot H_1(m) \mod n$ and sends this blinded message b_1 to KGC.

(4) **Signing**

- After receiving this blinded message b_1, KGC computes $b_2 = b_1^d \mod n$ and sends b_2 to each signer.
- For each $i = 1, \ldots, k$, P_i signs b_2 with his/her identity private key ppk_i to obtain $z_i = b_2 \cdot ppk_i \cdot x_i^{H_2(Y)}$ and sends z_i to the requester P_0.
- With z_i, $i = 1, \ldots, k$ in hand, the requester P_0 computes $Z = \Pi_{i=1}^k z_i$.

(5) **Extraction and verification**

- Upon receiving Z, the requester P_0 acquires the signature by computing

$$F = Z \cdot a^{-k} \mod n. \tag{1}$$

- The tuple (F, Y) constitutes a signature for m. And the correctness of this signature can be checked through

$$F^e \equiv H_1(m)^k \cdot Y^{H_2(Y)} \cdot \Pi_{i=1}^k H_0(ID_i) \mod n. \tag{2}$$

Next, we will show the correctness of our proposed scheme above. For any message m, the signature (F, Y) produced by our proposed scheme satisfies (2).

Theorem 1 *For any message m, if (F, Y) is the signature of m produced by the blind multi-signature scheme above, we have*

$$F^e \equiv H_1(m)^k \cdot Y^{H_2(Y)} \cdot \Pi_{i=1}^k H_0(ID_i) \mod n. \tag{3}$$

Proof This can be proved by the following equation:

$$\begin{aligned} F^e &\equiv \left(Z \cdot a^{-k}\right)^e \\ &\equiv \left(\Pi_{i=1}^k z_i \cdot a^{-k}\right)^e \\ &\equiv \left(\Pi_{i=1}^k \left(b_2 \cdot ppk_i \cdot x_i^{H_2(Y)}\right) \cdot a^{-k}\right)^e \\ &\equiv \left(\Pi_{i=1}^k \left(b_1^d \cdot x_i^{-1} \cdot sgn_i \cdot x_i^{H_2(Y)}\right) \cdot a^{-k}\right)^e \\ &\equiv \left(\Pi_{i=1}^k \left((a^e \cdot H_1(m))^d \cdot H_0(ID_i)^d \cdot x_i^{H_2(Y)}\right) \cdot a^{-k}\right)^e \\ &\equiv H_1(m)^k \cdot \left(\Pi_{i=1}^k \left(H_0(ID_i)^d \cdot x_i^{H_2(Y)}\right)\right)^e \\ &\equiv H_1(m)^k \cdot Y^{H_2(Y)} \cdot \Pi_{i=1}^k H_0(ID_i) \mod n. \end{aligned} \tag{4}$$

This scheme not only enables blind signature but also enables signers to sign based on their respective identity private keys. Furthermore, there is no need for secure channels to send the identity private keys, thereby cutting down on communication cost.

3 Security

We prove the security of our proposed scheme in this section. Let us first prove the following lemma.

Lemma 1 *Let* (n, e, d) *be the tuple of RSA encryption algorithm. For each* $t, u \in \mathbb{Z}_n^*$, $t \neq u$, *and* $a \in \mathbb{Z}$, *if* $t^a \equiv u^e \mod n$, *and* $\gcd(a, e) = 1$, *and then we can find* $x \in \mathbb{Z}_n^*$, *such that*

$$t \equiv x^e \mod n. \quad (5)$$

Proof Since $\gcd(a, e) = 1$, using the extended Euclidean algorithm, we can find $\alpha, \beta \in \mathbb{Z}$, such that

$$\alpha a + \beta e = 1. \quad (6)$$

We then have that

$$t \equiv t^{\alpha a + \beta e} \equiv (u^\alpha \cdot t^\beta)^e \mod n, \quad (7)$$

that is, we find $x = u^\alpha \cdot t^\beta$, such that

$$t \equiv x^e \mod n. \quad (8)$$

Now, we prove the security of our proposed scheme as follows.

Theorem 2 *Under the strong RSA assumption and the random oracle model, the blind multi-signature scheme introduced above can resist the existential forgery.*

Proof Assuming that an adversary $\mathcal{A}$ can break our multi-blind signature scheme, then according to the provable security method, there exists an algorithm $\mathcal{B}$ that can successfully solve the RSA one-way problem, i.e., for any $t \in \mathbb{Z}_m^*$, one can solve r satisfying $t \equiv r^e \mod n$, where (n, e, d) is the tuple of RSA encryption algorithm.

Algorithm $\mathcal{B}$ simulates the interaction between KGC and $\mathcal{A}$ and answers $\mathcal{A}$'s oracle queries by maintaining the corresponding lists.

(1) RSA one-way problem: Algorithm $\mathcal{B}$ publishes KGC's public key (n, e) and maintains initially empty lists $\{L_0, L_1, L_2\}$ associating with $\mathcal{A}$'s oracle queries.
(2) $H_0(ID_i)$ query: Algorithm $\mathcal{B}$ chooses randomly a number $ppk_i \in \mathbb{Z}_n^*$ and tosses a biased $coin_i$ which satisfies $coin_i = 0$ with probability θ, $\frac{1}{2} < \theta < 1$, and $coin_i = 1$ with probability $1 - \theta$. If $coin_i = 0$, then $\mathcal{B}$ computes $H_{0,i} = ppk_i^e \mod n$; if $coin_i = 1$, then it computes $H_{0,i} = t \cdot ppk_i^e \mod n$. Algorithm $\mathcal{B}$

stores tuple $(ID_i, ppk_i, coin_i, H_{0,i})$ in list L_0 and returns $H_{0,i}$ as $H_0(ID_i)$ to $\mathcal{A}$.

$H_1(m_i)$ query: Algorithm $\mathcal{B}$ looks up $(m_i, H_{1,i})$ in list L_1. If $(m_i, H_{1,i})$ exists in list L_1, it returns $H_{1,i}$ as $H_1(m_i)$ to $\mathcal{A}$. If this entry is not yet defined, it chooses randomly a number $H_{1,i} \in \mathbb{Z}_n^*$, stores tuple $(m_i, H_{1,i})$ in list L_1, and returns $H_{1,i}$ as $H_1(m_i)$ to $\mathcal{A}$.

$H_2(Y_i)$ query: Algorithm $\mathcal{B}$ looks up $(Y_i, H_{2,i})$ in list L_2. If $(Y_i, H_{2,i})$ exists in list L_2, it returns $H_{2,i}$ as $H_2(Y_i)$ to $\mathcal{A}$. If this entry is not yet defined, it chooses randomly a number $H_{2,i} \in \{0, 1\}^l$, stores tuple $(Y_i, H_{2,i})$ in list L_2, and returns $H_{2,i}$ as $H_2(Y_i)$ to $\mathcal{A}$.

(3) Key derivation query for P_i: Algorithm $\mathcal{B}$ looks up $(ID_i, ppk_i, coin_i, H_{0,i})$ in list L_0. If $coin_i = 0$, then $\mathcal{B}$ returns ppk_i to $\mathcal{A}$; otherwise, it exits execution.

(4) Signature query for $(m_i, ID_{i,1}, ID_{i,2}, \ldots, ID_{i,k})$: Algorithm $\mathcal{B}$ chooses randomly two numbers $F_i \in \mathbb{Z}_n$, $Y_i \in \mathbb{Z}_n^*$ and solves $H_{2,i}$ such that

$$F_i^e = H_{1,i}^k \cdot Y_i^{H_{2,i}} \cdot \Pi_{l=1}^k H_0(ID_{i,l}) \mod n. \tag{9}$$

If there are any entries in list L_2 that contain Y_i or $H_{2,i}$, then $\mathcal{B}$ reselects F_i, Y_i, and computes $H_{2,i}$ until Y_i, $H_{2,i}$ appear in list L_2 for the first time. $\mathcal{B}$ stores tuple $(Y_i, H_{2,i})$ in list L_2 and returns the signature (F_i, Y_i) to $\mathcal{A}$. It is clear that $\mathcal{A}$ can verify the correctness of this signature.

(5) Adversary $\mathcal{A}$ forgery: Eventually, $\mathcal{A}$ outputs a forged signature $(\tilde{F}, \tilde{Y})$ with identity set $U = \{ID_1, \ldots, ID_k\}$ and message $\tilde{m}$. The signature is nontrivial, that is, $\mathcal{A}$ has not queried U about the blind multi-signature of $\tilde{m}$, at least one of the identity private keys of signers in U has not been queried, and there is at least one signer in U whose value of $coin$ is equal to 1 in list L_0.

Let

$$U_1 := \{ID_i : ID_i \in U \text{ and } coin_i = 1\},\ U_0 := U \setminus U_1; \tag{10}$$

then

$$\begin{aligned}\tilde{F}^e &= H_1(\tilde{m})^k \cdot \tilde{Y}^{H_2(\tilde{Y})} \cdot \prod_{i \in U_1} H_0(ID_i) \cdot \prod_{i \in U_0} H_0(ID_i) \mod n \\ &= H_1(\tilde{m})^k \cdot \tilde{Y}^{H_2(\tilde{Y})} \cdot t^{|U_1|} \cdot \prod_{i \in U} ppk_i^e \mod n,\end{aligned} \tag{11}$$

that is,

$$t^{|U_1|} = \left(\frac{\tilde{F}}{\prod_{i \in U} ppk_i \cdot H_1(\tilde{m})^{k/e} \cdot \tilde{Y}^{H_2(\tilde{Y})/e}} \right)^e \mod n. \tag{12}$$

If $\gcd(e, |U_1|) = 1$, then by Lemma 1, we can find $\alpha, \beta \in \mathbb{Z}$ such that $\alpha e + \beta |U_1| = 1$ and

$$t = \left(t^{\alpha} \cdot \left(\frac{\tilde{F}}{\prod_{i \in U} ppk_i \cdot H_1(\tilde{m})^{k/e} \cdot \tilde{Y}^{H_2(\tilde{Y})/e}} \right)^{\beta} \right)^{e} \mod n. \tag{13}$$

Let

$$r = t^{\alpha} \cdot \left(\frac{\tilde{F}}{\prod_{i \in U} ppk_i \cdot H_1(\tilde{m})^{k/e} \cdot \tilde{Y}^{H_2(\tilde{Y})/e}} \right)^{\beta}; \tag{14}$$

then algorithm $\mathcal{A}$ can therefore output r as the RSA inversion of t.

According to Theorem 2, our proposed blind multi-signature scheme is secure.

4 Conclusion

Based on RSA cryptography, we introduce an efficient broadcast blind multi-signature scheme, which enables the production of a multi-signature with a length similar to that of a single signature, and the security of our scheme is proved under the RSA one-way problem in the random oracle model. By importing a key generation center, it works with all signers to generate their respective identity private keys based on a unique private key of RSA. It reduces the computational complexity of traditional key frameworks and avoids the dependence on secure channels during the distribution of identity private keys, reducing the communication cost of multi-signatures.

Acknowledgments We would like to acknowledge the assistance of anonymous reviewers in making their valuable comments which greatly improved this work. This work is supported by the China Postdoctoral Science Foundation (Grant No. 2024T170371) and Young Talent of Lifting Engineering for Science and Technology in Shandong, China (Grant No. SDAST2024QTA067).

References

1. Chaum, D.: Blind signatures systems. In: Advances in Cryptology-Proceedings of CRYPTO '83, Plenum, pp. 153 (1983)
2. Batten, L., Yi, X.: Off-line digital cash schemes providing untraceability, anonymity and change. Electron. Commer. Res. **19**, 81–110 (2019)

3. Islam, S., Amin, R., Biswas, G., et al.: Provably secure pairing-free identity-based partially blind signature scheme and its application in online e-cash system. Arab. J. Sci. Eng. **41**, 3163–3176 (2016)
4. Liu, J., Liu, J., Qiu, X.: A proxy blind signature scheme and an off-line electronic cash scheme. Wuhan Univ. J. Nat. Sci. **18**, 117–125 (2013)
5. Nguyen, T., Dang, T.: Enhanced security in internet voting protocol using blind signature and dynamic ballots. Electron. Commer. Res. **13**, 257–272 (2013)
6. Zhang, X., Zhang, J., Xie, S.: A secure quantum voting scheme based on quantum group blind signature. Int. J. Theor. Phys. **59**, 719–729 (2020)
7. Li, W., Shi, R., Guo, Y.: Blind quantum signature with blind quantum computation. Int. J. Theor. Phys. **56**, 1108–1115 (2017)
8. Zou, X., Qiu, D.: Attack and improvements of fair quantum blind signature schemes. Quantum. Inf. Process. **12**, 2071–2085 (2013)
9. Itakura, K., Nakamura, K.: A public-key cryptosystem suitable for digital multisignatures. NEC Res. Develop. **71**, 1–8 (1983)
10. Harn, L., Kresler, T.: New scheme for digital multisignatures. Electron. Lett. **25**, 1002–1003 (1989)
11. Bellare, M., Neven, G.: Identity-based multi-signatures from RSA. In: Topics in Cryptology - CT-RSA 2007, The Cryptographers Track at the RSA Conference 2007, San Francisco, February 2007, pp. 145–162 (2007)
12. Bagherzandi, A., Jarecki, S.: Identity-based aggregate and multi-signature schemes based on RSA. In: 13th International Conference on Practice and Theory in Public Key Cryptography, Paris, May 2010, pp. 480–498 (2010)
13. Tsai, J., Lo, N., Wu, T.: New identity-based sequential aggregate signature scheme from RSA. In: International Symposium on Biometrics and Security Technologies, Chengdu (2013)
14. Hohenberger, S., Waters, B.: Synchronized aggregate signatures from the RSA assumption. In: Advances in Cryptology - EUROCRYPT 2018 - 37th Annual International Conference on the Theory and Applications of Cryptographic Techniques, Tel Aviv, pp. 197–229 (2018)
15. Boneh, D., Lynn, B., Shacham, H.: Short signatures from the Weil pairing. In: Advances in Cryptology - ASIACRYPT 2001, 7th International Conference on the Theory and Application of Cryptology and Information Security, Gold Coast, Australia, pp. 514–532 (2001)
16. Chen, T., Huang, K., Chung, Y.: Digital multi-signature scheme based on the Elliptic Curve cryptosystem. J. Compt. Sci. Technol. **19**, 570–573 (2004)
17. Ristenpart, T., Yilek, S.: The power of proofs-of-possession: Securing multiparty signatures against rogue-key attacks. In: Advances in Cryptology - EUROCRYPT 2007, 26th Annual International Conference on the Theory and Applications of Cryptographic Techniques, Barcelona, pp. 228–245 (2007)
18. He, D., Zeadally, S., Xu, B., et al.: An efficient identity-based conditional privacy-preserving authentication scheme for vehicular ad hoc networks. IEEE Trans. Inf. Forens. Secur. **10**, 2681–2691 (2015)
19. Boneh, D., Drijvers, M., Neven, G.: Compact multi-signatures for smaller blockchains. In: Advances in Cryptology - ASIACRYPT 2018 - 24th International Conference on the Theory and Application of Cryptology and Information Security, Brisbane, December 2018, pp. 435–464 (2018)
20. Petersen, H., Horster, P., Michels, M.: Blind multisignature schemes and their relevance to electronic voting. In: 11th Annual Computer Security Applications Conference, IEEE Press, pp. 149–155 (1995)
21. Chen, F., Hu, M., Sun, A.: Sequential blind multisignature based on hyperelliptic curve. Comput. Eng. **37**, 160–162 (2011)
22. Jiang, A., Du, W.: ElGamal type broadcasting multi-blind signature scheme based on elliptic curve. Comput. Appl. Softw. **27**, 287–290 (2010)
23. NIST: Federal Information Processing Standards Publication 180-4, Secure Hash Standards (SHS), National Institute of Standards and Technology, pp. 1–36 (2015)

Similar Word Replacement: A New Adversarial Example Generation Strategy for Chinese Text Classification

Fangzhou Yuan, Hongjun Wang, Bingchuan Wang, Yuan Yuan, and Mingfeng Lu

1 Introduction

Deep learning has developed rapidly in the past 10 years. It has promoted breakthroughs in many fields, such as computer vision, speech processing, and natural language processing (NLP). Many classification models are used for text classification tasks, and they have achieved good results.

In 2015, the concept of adversarial examples was proposed [1]. After adding subtle disturbances to the input of the deep neural network model, the prediction results can be easily changed with high confidence. The adversarial example phenomenon demonstrates the security problems of deep learning-based artificial intelligence models. Since the deep learning model is a black box with insufficient interpretability, unexpected security risks may also occur without malicious attacks. Due to the lack of robustness of this classification model, the use of AI technology in highly security-sensitive scenarios, such as medical care, transportation, and the military, would be affected.

To ensure that the text adversarial examples are still readable, the modification strategy needs to be different from adding specific noise. In addition to not affecting the original intent of the text and being able to deceive the classifier,

F. Yuan
Luoyang, China

School of Information and Electronics, Beijing Institute of Technology, Beijing, China

H. Wang · B. Wang · Y. Yuan
Luoyang, China

M. Lu (✉)
School of Integrated Circuit and Electronics, Beijing Institute of Technology, Beijing, China
e-mail: lumingfeng@bit.edu.cn

X. Jia et al. (eds.), *2nd International Conference on Big Data, IoT, and Cloud Computing*, EAI/Springer Innovations in Communication and Computing,
https://doi.org/10.1007/978-3-031-93570-1_22

Table 1 English letter-level modification methods

Modification method	Target word	Modified word
Insertion	Language	Languaoge
Deletion		Laguage
Exchange		Langugae
Replacement		Larguage

Table 2 Modification methods for Chinese text classification adversarial example generation methods

Modification method	Target word	Modified word
Homophone replacement	欢迎	幻影
Traditional character replacement		迎
Pinyin rewriting		Huanying
Word disassembly		欢，迎
Word order disturbance		迎欢

a high-quality text classification adversarial example should meet the following extra conditions: first, human observers should not easily find the modification; second, human observers should not easily conclude the modification strategy. At present, the modification strategies of text classification adversarial example generation algorithms are mostly designed around these conditions. There are many adversarial example generation methods for English text classification algorithms, such as HotFlip, DeepWordBug, Textbugger, TextFooler, PWWS, and improved DeepWordBug [2–12].

English words are made of letters, so they can be modified slightly at the letter level. Table 1 shows some examples of letter-level modification methods. Chinese words and texts are composed of Chinese characters. Different from letter-level modification in English, modification of any Chinese character will be conspicuous in the text and easy to be discovered by human observers. Due to the difference between Chinese and English, the modification methods of the Chinese text are quite different. In recent years, the research on Chinese text classification adversarial examples is increasing. Algorithms, such as WordHandling, CWordAttacker, Marginal Attack, WordChange, Argot, WordHit, WordErrorSim, and PGAS, have achieved good results in deceiving Chinese text classifiers [13–27]. Most of these algorithms use modification strategies such as homophone replacement, traditional character replacement, Pinyin rewriting, phrase disassembly, word order disturbance, and character disassembly. Some of the modification methods are shown in Table 2.

Most of these algorithms have paid more attention to the success rate of adversarial attacks and used Word Mover's Distance (WMV) [28] to measure text similarity. This brings about some problems. First, the modification methods are very obvious, and human observers can easily find the changes and summarize the modification rules. Second, text similarity based on WMV cannot accurately measure the deception of the adversarial examples to human observers. This is because the subversive modification of a keyword in the text (e.g., changing a positive emotion word into a negative emotion word) can make the text classifier misclassify with a small WMV. However, this modification also changes the original meaning of the text, and it is easy for human observers to find it.

To ensure that the adversarial examples can make the classifier misclassify and reduce the discovery probability of the changes and modification strategies, we proposed a new Chinese text classification adversarial example generation strategy: similar word replacement (SWR). The main contributions of this paper are as follows:

1. We proposed a Chinese adversarial example generation strategy that successfully deceives both the RNN- and CNN-based classifiers and human observers.
2. We divided Chinese words into semantically similar and appearance-similar words and proposed the generation method, respectively. The generation method of appearance-similar words comprehensively considers four factors: Chinese character's image, Fivepen code, Zhengma code, and stroke order code, which is a brand new method.
3. We comprehensively used semantically similar word replacement, appearance-SWR, traditional character replacement, and word order disturbance to form a new Chinese text adversarial example generation strategy and named it SWR.
4. We proposed an evaluation method for adversarial examples' quality. This method is a subjective evaluation method, which comprehensively considers three subjective indicators: original intention retention, modification concealment, and strategy concealment.
5. Since the deceptiveness of human observers is considered, we formed an evaluation group consisting of 16 people proficient in Chinese to evaluate human observers. Experimental results and human observer test results showed that SWR is an effective adversarial example generation strategy for Chinese text classification. Compared with other methods, it not only performs well in the classification accuracy but also achieves higher scores than other methods in all three indicators in the human observer test.

2 Chinese Similar Words

A Chinese similar word is a Chinese word similar to the original Chinese word. The Chinese text classification adversarial example generation method based on SWR proposed in this paper is to find a similar word to the target Chinese word to replace it. Since the similar word and the target word have different effects on the text classification results, when one of the similar words affects the real classification result, the target word is replaced to generate an adversarial example. SWR is hardly noticed by human observers.

Chinese similar words are divided into two types: semantically similar and appearance-similar words.

Table 3 Examples of the search and calculation results for semantically similar words

Target word	Semantically similar words	The highest similarity value
傲慢	高傲, 自傲, 蔑视, 自大, 盛气凌人, 倨傲, 狂妄, 目中无人	高傲 0.819
白日梦	美梦, 幻想, 空想, 梦, 妄想, 梦境, 做梦, 不切实际	美梦 0.756
开开心心	快快乐乐, 高高兴兴, 开心快乐, 幸福快乐, 开心, 欢欢喜喜, 健健康康, 平平安安	快快乐乐 0.878

2.1 Semantically Similar Words

The concept of semantically similar words is similar to synonyms in Chinese, such as "高兴"and "开心", and "伤心"and "难过". Through the word vector, the similarity between words can be calculated, and the semantically similar word of a specific Chinese word can be searched. Tencent Chinese word vector is used in this paper to search for semantically similar words. In 2018, Tencent Chinese word vector became open-source by Tencent AILab. This Chinese word vector contains more than eight million Chinese words, and each Chinese word corresponds to a 200-dimensional vector. Compared with other Chinese word vectors, this word vector has greatly improved in terms of freshness, coverage, and other indicators.

The similarity value is the cosine similarity calculated based on the 200-dimensional vector of the two words, ranging from 0 to 1; the closer to 1, the higher the similarity. Table 3 shows some examples of the search results; eight words with the highest similarity value are the output.

2.2 Appearance-Similar Words

Characters are stored in the computer as a piece of code, but for human observers, a character or letter is a special figure. In particular, Chinese characters, which evolved from hieroglyphs, are stored graphically in the human brain. If two Chinese characters are very similar in appearance, human observers will easily confuse them. Appearance-similar characters can be used to generate appearance-similar words.

Researchers have used Chinese character structure, outline, stroke, writing order, and other characteristics to search for appearance-similar characters. They proposed a font similarity algorithm based on character image and stroke order code [29]. However, sometimes too much consideration of the stroke order cannot help us get the most similar character. There are also problems in searching for appearance-similar characters by calculating the difference between two Chinese character images. Some characters with the same radical look very similar, but in the image, the radical occupies different pixel positions.

2.3 *The Searching Algorithm for Appearance-Similar Characters*

The Fivepen and Zhengma input methods are both commonly used Chinese input methods based on the structure of Chinese characters. They disassemble Chinese characters into roots. The difference is that the Fivepen input method uses "character pieces" to group characters, while the Zhengma input method disassembles Chinese characters into basic roots according to the writing order. Both of them can reflect the structural characteristics of the Chinese character, so they are chosen to search appearance-similar character.

A new searching algorithm based on the literature [29] for appearance-similar characters is proposed, which comprehensively consider the appearance of Chinese character image, Fivepen code, Zhengma code, and stroke order. It is designed as follows:

1. Acquire Chinese character information.
   Traverse the standard Chinese character database, and obtain the corresponding pictures, the Fivepen code, the Zhengma code, and the stroke order code for all Chinese characters.
2. Calculate the difference between images, Fivepen codes, Zhengma codes, and stroke order codes between two Chinese characters.
   First, calculate the difference dif_1 between two Chinese characters' images:

$$dif_1 = sum\left(|A - B|\right)/S \tag{1}$$

A and B are the binary matrix of the two Chinese characters' images, S is the number of pixels in the black part of the target Chinese character's image, and $sum(|A-B|)$ can be understood as the "area" of the non-overlapping part of two Chinese characters. Then, the relative difference between the Chinese characters can be calculated, uniformly quantified, and compared.

Second, same as in [29], based on the Jaro-Winkler distance of the Fivepen code, calculate the similarity sim_2 and then calculate the difference between two Chinese characters' Fivepen code dif_2:

$$sim_2 = dis_2 + len_{AB} \cdot \left[\frac{1 - dis_2}{\text{Max}\left(len_A, len_B\right)}\right] \tag{2}$$

$$dif_2 = 1 - sim_2 \tag{3}$$

Table 4 Some search results of appearance-similar characters

Target character	Appearance-similar character	Difference value
汩	汨	0.022727273
菓	菒	0.029411765
苻	茌	0.899871
莝	茲	0.9
秫	秼	0.9
鍩	鋯	0.900068
頕	頔	0.900154

The calculation method of the difference between two Chinese characters' Zhengma code and stroke order code dif_3 and dif_4 are the same as dif_2.

3. Calculate the difference between two Chinese characters.
   Calculate the difference between two Chinese characters *Dif*:

$$Dif = \omega_1 dif_1 + \omega_2 dif_2 + \omega_3 dif_3 + \omega_4 dif_4 \tag{4}$$

ω_1, ω_2, ω_3, and ω_4 are the weight values of the difference between images, Fivepen codes, Zhengma codes, and stroke order codes. Small-scale experiment results show that when the four weights are the same, the calculated similar Chinese characters are closer to the results of the human test. The examples of the algorithm running results are shown in Table 4.

2.4 *The Generation Algorithm for Appearance-Similar Words*

There are a total of 20,902 Chinese characters in the standard Chinese character library. Not all of them have another Chinese character that looks very similar. As shown in Table 4, some search results are not similar to the target Chinese characters. To ensure that each target word has an available modification strategy, traditional character replacement and word order disturbance are added to the modification strategies to generate appearance-similar words. Traditional character replacement is to replace every character in the target word with its traditional form. Word order disturbance is based on the conclusion that "the change of word order does not affect reading to a certain extent [30]."

The specific steps for generating appearance-similar words are as follows:

1. Determine the modification strategy.
   Appearance-similar character replacement: Split the target word into independent Chinese characters. For each character, the appearance-similar character searching algorithm is used. If the difference value of the appearance-similar

Table 5 Examples of appearance-similar words

Target word	Appearance-similar words
汨罗	汩罗,汨羅, 罗汨
口头禅	口头禪,口頭禅,口头禪,口禅头
机器学习	杋器学习, 机嚚学习, 機器学习,机器學习,机器学習,机器习学

character is less than threshold A, the target character is replaced, and an appearance-similar word is generated. In this paper, the threshold A is set to 0.9.

Traditional character replacement: Split the target word into independent Chinese characters. For each character, get its traditional form. If the traditional form is different, the target character is replaced, and an appearance-similar word is generated.

Word order disturbance: If the target word has more than one character, the last two characters in the target word are exchanged to generate an appearance-similar word.

2. Generate a group of appearance-similar words.
   Modify the target word according to the modification strategy, and a group of appearance-similar words are generated. The results of the generated appearance-similar words for some target words are shown in Table 5.

3 The Adversarial Example Generation Algorithm Based on SWR

Based on the generation algorithm for Chinese similar words, the generation algorithm of adversarial example for Chinese text classification based on SWR is proposed. The main idea is to determine multiple modification strategies based on SWR, and the most effective strategy is chosen according to the confidence of the output on the original or target category.

3.1 Target Text Preprocessing

Remove the punctuation, numbers, letters, symbols, and common words of the target text X, divide the text into Chinese words, and then a list $T = (x_1, x_2, x_3, \ldots, x_m)$ containing m words is generated. The purpose of target text preprocessing is to decompose the target Chinese text into Chinese words.

3.2 The Importance Ranking of the Target Text Words

The m words in the word list T are deleted in turn, and a new text is generated every time. These new texts with a single word deleted are the input to the classifier F, and confidence in the original category or target category can be obtained. The confidence value is used as the importance score of the current deleted word. The calculation formula for the word importance score is as follows:

$$C_i = F_y\left(x_1, x_2, \ldots x_{i-1}, x_{i+1}, \ldots x_{m-1}, x_m\right) \tag{5}$$

In a non-target attack, C_i represents the confidence of the new text with the i-th word deleted on the original category y. The lower the C_i, the lower the output confidence of the new text, which proves that modifying x_i is more beneficial for classifying to other categories, and confidence scores are sorted in ascending order. Oppositely, in a target attack, C_i represents the confidence in the target category y, and the confidence scores are sorted in descending order. The higher the C_i, the higher the output confidence of the new text, which proves that modifying x_i is more beneficial for classifying to the target category y, and confidence scores are sorted in ascending order. After the importance ranking, the text can be modified in this order to generate an adversarial example.

3.3 SWR

According to the importance ranking, the operation of generating similar words is performed on the i-th word x_i, and n similar words x_{i1}' ~ x_{in}' are obtained. By replacing the target word x_i, n new texts X_1' ~ X_n' can be generated.

In a non-target attack, X_1' ~ X_n' are the input into the classifier F, and confidence in the original category is the output. Then, select the text with the lowest confidence. If the confidence value is lower than that of the target text X, the modification is effective, and a new text X' is generated. If the confidence value is not lower than that of the target text X, the modification is invalid, and the $i + 1$-th word needs to be modified. The specific calculation process is shown in Formulas (6), (7), and (8).

$$X'_k = \left(x_1, x_2, \ldots x_{i-1}, x_{ik}{}', x_{i+1}, \ldots x_{m-1}, x_m\right) \tag{6}$$

$$X^* = argmin\left(F_y\left(X'_1\right), F_y\left(X'_2\right), \ldots F_y\left(X'_n\right),\right) \tag{7}$$

$$\begin{cases} X' = X^* & \left(F_y\left(X^*\right) < F_y(X)\right) \\ i = i + 1 & \left(F_y\left(X^*\right) \geq F_y(X)\right) \end{cases} \tag{8}$$

In a target attack, X_1' ~ X_n' are the input into the classifier F, and confidence in the target category is the output. Then, select the text with the highest confidence. If the confidence value is higher than that of the target text X, the modification is effective, and a new text X' is generated. If the confidence value is not higher than that of the target text X, the modification is invalid, and the $i + 1$-th word needs to be modified. The specific calculation process is shown in formulas (6), (9), and (10).

$$X^* = argmax\left(F_y\left(X_1'\right), F_y\left(X_2'\right), \ldots F_y\left(X_n'\right),\right) \tag{9}$$

$$\begin{cases} X' = X^* & \left(F_y\left(X^*\right) > F_y(X)\right) \\ i = i + 1 & \left(F_y\left(X^*\right) \leq F_y(X)\right) \end{cases} \tag{10}$$

3.4 Effectiveness Judgment

In a non-target attack, if the confidence value of X' in the original category is not the maximum in m categories, it means that the classification result is not the original category, and X' could be the output as the adversarial example X_{AE}. Oppositely, if the confidence value is still the maximum in all categories, it means that the classification result is still the original category.

$$\begin{cases} X_{AE} = X' & \left(F_y\left(X'\right) \neq \max\left(F_1\left(X'\right), \ldots F_m\left(X'\right)\right)\right) \\ j = j + 1 & \left(F_y\left(X'\right) = \max\left(F_1\left(X'\right), \ldots F_m\left(X'\right)\right)\right) \end{cases} \tag{11}$$

In a target attack, if the confidence value of X' in the target category is the maximum in m categories, it means that the classification result is the target category, and X' could be the output as the adversarial example X_{AE}. Oppositely, if the confidence value is not the maximum in all categories, it means that the classification result is not the target category.

$$\begin{cases} X_{AE} = X' & \left(F_y\left(X'\right) = \max\left(F_1\left(X'\right), \ldots F_m\left(X'\right)\right)\right) \\ j = j + 1 & \left(F_y\left(X'\right) \neq \max\left(F_1\left(X'\right), \ldots F_m\left(X'\right)\right)\right) \end{cases} \tag{12}$$

If X' could not pass the judgment, one more word in X' needs to be modified to generate the adversarial example. To ensure the quality of adversarial examples, a threshold N should be set for the number of modifications. If the number of modifications j has exceeded N, the generation of the adversarial example of the text fails.

4 Experimental Results and Analysis

4.1 Datasets and Target Classification Models

Two datasets are selected for the experiment on the adversarial example generation algorithm based on SWR, which are the weibo_senti_100k sentiment classification dataset and the THUCNews news classification dataset.

The weibo_senti_100k sentiment classification dataset is a two-class classification dataset; the average length is 42.9 words. This dataset contains 100,000 pieces of Sina Weibo text data, including two categories: positive emotion and negative emotion. Each category has 50,000 pieces of data.

The THUCNews news classification dataset contains 50,000 pieces of data; the average length is 534.53 words, including 10 categories: finance, real estate, home furnishing, education, technology, fashion, political events, sports, games, and entertainment. Each category has 5000 pieces of data.

The two datasets are different in length and the number of categories. The purpose is to test whether the algorithm can achieve good results on the short text and long text, multi-classification and two-class classification.

Two kinds of classification models are chosen as the target classification model: LSTM (based on RNN) and TextCNN (based on CNN). They were trained, respectively, and the classification accuracy on the two datasets is shown in Tables 6 and 7.

In weibo_senti_100k, 2000 pieces of data from the test set are selected to form an experimental dataset, with 1000 pieces of data for each category. In THUCNews, 100 pieces of data from the test set are selected to form an experimental dataset, with 10 pieces of data for each category. The classification results of the experimental dataset are shown in Table 8.

Table 6 Classification results of weibo_senti_100k

Classification model	Accuracy	Precision	Recall	f1-score
LSTM	95.6%	94.0%	97.1%	0.956
TextCNN	92.4%	95.4%	90.0%	0.926

Table 7 Classification results of THUCNews

Classification model	Accuracy	Precision	Recall	f1-score
LSTM	95.2%	96.4%	96.6%	0.965
TextCNN	96.1%	97.4%	95.9%	0.966

Table 8 Classification result of the experimental dataset

Dataset	Data volume	Classification accuracy
weibo_senti_100k	2000	LSTM: 96.3% TextCNN: 93.2%
THUCNews	100	LSTM: 100% TextCNN: 100%

4.2 Experimental Results

In this experiment, three kinds of strategies are used to generate an adversarial example: semantically similar word replacement, appearance-similar word replacement, and SWR, and the generated adversarial examples are used to attack the two target models. Since weibo_senti_100k is a two-class classification dataset, this experiment did not distinguish between a target attack and a non-target attack. The threshold N of the maximum number of modified words is determined by the average length of the dataset. The maximum modification ratio of the text is set to 20%. In weibo_senti_100k, the average length is 42.9 words, the threshold $N = 42.9 \times 0.2 = 8.58$, and N is 8. In THUCNews, the average length is 534.53 words, the threshold $N = 534.5 \times 0.2 = 106.9$, and N is 100.

Figure 1 shows the experimental results. With the increase of number *N*, the classification accuracy of the classifier keeps declining, which indicates that the modification ratio of the text will significantly affect the effect of the adversarial attack. Compared to semantically similar word replacement and appearance-similar word replacement, the success rate of SWR has been greatly improved. The classification accuracy rate has reduced from 40% to 20% in weibo_senti_100k, from 40% to 30% in THUCNews. It proves that SWR has achieved obvious attack effects as a combination of two modification strategies. In weibo_senti_100k, the classification accuracy is not much different from the previous two strategies when $N = 1$, but it is much lower when $N \geq 2$. That is because when $N = 1$, the modification of the text is limited to one time, and only the relatively better one can be selected from the two strategies. When $N \geq 2$, the algorithm will choose a better strategy according to the feedback of the confidence, which can reflect the advantages of SWR.

Tables 9, 10, 11, and 12 show some of the adversarial examples and their original texts. Table 13 shows the modification strategies used in the examples. The generated adversarial examples could deceive human observers successfully, and without the modified words marked, it is hard to find the modification. It can be seen that SWR is used in the adversarial examples, including semantically similar word replacement and appearance-similar word replacement. The strategies to generate appearance-similar words, such as appearance-similar character replacement, traditional character replacement, and word order disturbance, could all be found. SWR is a combination of multiple strategies, which is beneficial to protect itself from being discovered. If human observers know that the text is an adversarial example, it is also very hard for them to find out or conclude the modification strategy.

To prove the advantages of our algorithm over existing algorithms, we select DeepWordBug [7] and WordHandling [13] as the baseline methods in the experiment. Since WordHandling is a binary text classification algorithm, only the experimental dataset of weibo_senti_100k is used for the experiment. The experimental conditions are the same as above. LSTM and TextCNN are the targets of adversarial attack, and the maximum modification ratio of the text is set to 20%.

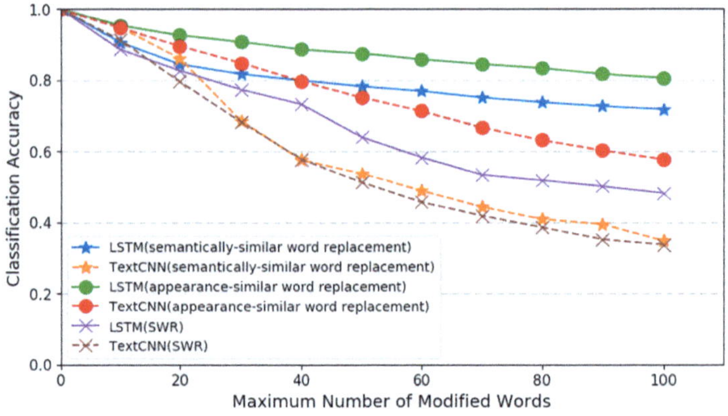

a

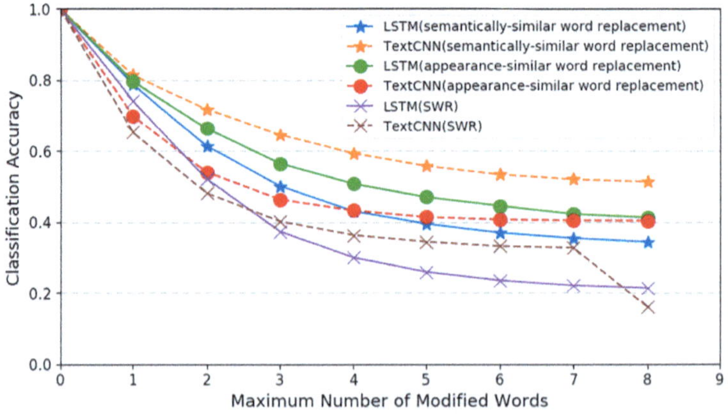

b

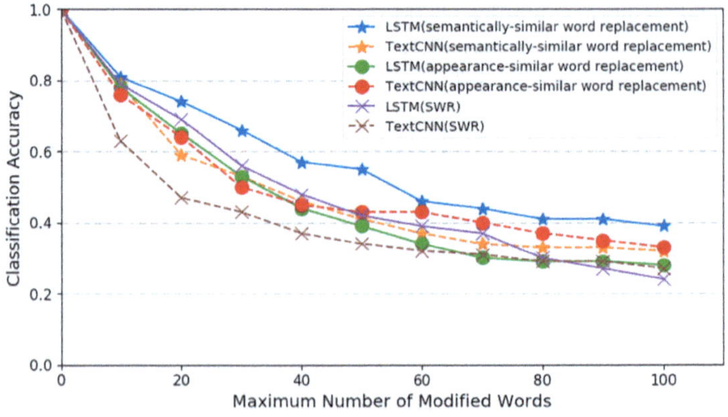

c

Fig. 1 Experimental results (**a**) the weibo_senti_100k experimental result, (**b**) the THUCNews experimental result (target attack), (**c**) the THUCNews experimental result (non-target attack)

Table 9 Examples of adversarial examples and their original texts on TextCNN in weibo_senti_100k

Original Text	Adversarial Example
[哈哈]//@心中一股冲劲勇闯: 癫狂了 (positive emotion)	[哈哈]//@心中一股冲劲勇闯: 癫狂了 (negative emotion)
早就是这样了。周六的书评周刊对行业影响很大!//@陶然万柳世界:[衰]《新京报》是我家订的唯一一份报纸，早知这样今年不订了 (negative emotion)	早就是这样了。周六的书评周刊对行业影响很大!//@陶然万柳界世:[衰]《新京报》是我家订的唯一一份报纸，只道这样今年不订了 (positive emotion)
感谢爸妈 ~ ~ ~ [爱你] (positive emotion)	谢谢你们爸妈 ~ ~ ~ [爱你] (negative emotion)

Table 10 Examples of adversarial examples and their original texts on LSTM in weibo_senti_100k

Original Text	Adversarial Example
喝多都不忘又吃又拿的，靠谱[哈哈] (positive emotion)	断片都不忘又吃又拿的，靠谱[哈哈] (negative emotion)
我必须强迫自己先厥过去，不然肯定走远了[衰] //@蕉蕉卉籽:@潘晓婷[嘻嘻]死不死?(negative emotion)	我必须强自己先厥过去，不然肯定走远了[衰] //@蕉蕉卉籽:@潘晓婷[嘻嘻]死不死?(positive emotion)
害怕着同时惊艳啊//@Ben9号:去了就回不来了! //@淡若天涯张小爱:@Ben9号 蓝洞，蓝洞，我们那未去成的帕劳蓝洞![泪][泪][泪] (negative emotion)	害怕着同时惊艳啊//@Ben9号:去了就回不了! //@淡若天涯张小爱:@Ben9号 蓝洞，蓝洞，我们那未去成的帕劳蓝洞![泪][泪][泪] (positive emotion)

Table 11 Examples of adversarial examples and their original texts on LSTM in THUCNews

Original Text	Adversarial Example
全球家具建材CBD 即将正式落户成都...南方香江集团共同打造的超级批发集散商务平台——“全球家具建材CBD”即将正式落户成都...承接沿海和国家优势产业资源的转移，带动本地产业升级。 (home furnishing)	全球家具建材CBD 即将正式落户成都...南方香江集团共同打造的超级批发集散商业人士平台——“全球家具建材CBD”即将正式落户成都...承接沿海和国家势产业资源的迁移，带动本地产业升级。 (games)

Table 12 Examples of adversarial examples and their original texts on LSTM in THUCNews

Original Text	Adversarial Example
卫生部:三年内实现农民在省内看病即时报销...(记者刘天思) ...卫生部今天下发通知，要求省、市级新农合定点医疗机构...卫生部农村卫生管理司副司长聂春雷说...因为过去他是在省级和市级医疗机构看病...就是新农合报的钱就不用自己出了...通过定点医疗机构现场结报就能够有效地杜绝这样的现象发生。” (political events)	卫生部:三年内实现农民在省内看病即时报销...(北青报刘天思) ...卫生部今天下发通知，严格要求省、市级新农合定点医疗机构...卫生部农村卫生管理司副局长聂春雷说...因为过去他是在全省和市级医疗机构看病...就是新农合报的钱就不需要自己出了...通过定点医疗机构现场照片结报就能够有效地杜绝这样的现象发生。” (education)

Table 13 Modification strategies used in the examples

Modification strategy		Target word	Modified word
Semantically similar word replacement		早知	只道
		感谢	谢谢你们
		喝多	断片
		商务	商业人士
		转移	迁移
		记者	北青报
		要求	严格要求
		司长	局长
		省级	全省
		不用	不需要
		现场	现场照片
Appearance-similar word replacement	Appearance-similar character replacement	强迫	强廹
		来	來
		泪	汨
		冲劲	優势
		管理	菅理
	Traditional character replacement	爱你	愛你
		优势	優势
	Word order disturbance	世界	界世

Table 14 Experimental results of baseline methods

Classification model	Original Accuracy	DeepWordBug	WordHandling	SWR
LSTM	96.3%	46.3%	37.1%	21.5%
TextCNN	93.2%	43.4%	36.7%	16%

Table 14 shows that the SWR-based adversarial attack achieves the largest reduction in classification accuracy. Compared with the two existing algorithms, the adversarial example generation strategy, SWR, is more effective in deceiving the Chinese text classifier.

4.3 Human Observer Test

The evaluation group consisted of 16 members of college teachers or students, born and raised in China, and who can speak and use Chinese proficiently.

First, a set of adversarial examples generated by the SWR strategy is mixed with several normal texts to test whether the evaluation group can find anomalies and classify the texts correctly. The test result shows that 87.5% (14/16) of the evaluation group did not find the modification in the set of texts, and 100% (16/16) made the

Table 15 The subjective indicators scoring results of adversarial examples

Modification strategy	Original intention retention scoring	Modification concealment scoring	Strategy concealment scoring
Homophone replacement	78.75	63.75	61.25
Pinyin rewriting	76.56	48.75	45.31
Character disassembly	81.88	50.63	47.19
Word disassembly	81.56	65.31	57.5
DeepWordBug	72.19	64.75	55.69
WordHandling	80.38	60.31	47.06
SWR	87.73	76.56	73.91

correct classification. This means that the modification of the adversarial examples generated by this strategy is sufficiently hidden from human observers, and it is difficult for human observers to perceive the modification.

Second, a set of adversarial examples generated by the SWR strategy is mixed with adversarial examples generated by other strategies, such as homophone replacement, Pinyin rewriting, word disassembly, character disassembly, DeepWordBug, and WordHandling. Since original intention retention, modification concealment, and strategy concealment are all important factors reflecting the quality of an adversarial example, they are chosen as subjective indicators and scored by the evaluation group. The scoring result is shown in Table 15. It can be seen that the adversarial examples generated by SWR performed better than other modification strategies in the three factors, especially in modification concealment and strategy concealment.

4.4 *The Transferability of the Algorithm*

The transferability of adversarial examples refers to the ability of adversarial examples generated on one classification model to deceive another classification model. In this experiment, the adversarial examples generated by the LSTM classification model are used to attack the TextCNN classification model, and the adversarial examples generated by the TextCNN classification model are used to attack the LSTM classification model. Both generation methods are based on SWR. The experimental results are shown in Table 16.

It can be seen that the attack success rate is not ideal. Concerning weibo_senti_100k, it is barely able to reach a success rate of about 30%, while for THUCNews, it is hard to complete the attack. Perhaps this is because our algorithm is a gray box attack algorithm, which needs the output confidence of the target classifier as a basis to constantly iteratively modify the text.

Table 16 The experimental results of transferability

Original model	Target model	Dataset	Success rate
TextCNN	LSTM	weibo_senti_100k	26.3%
		THUCNews(target attack)	2.04%
		THUCNews(non-target attack)	3.7%
LSTM	TextCNN	weibo_senti_100k	36.2%
		THUCNews(target attack)	7.51%
		THUCNews(non-target attack)	15.3%

5 Conclusion

In this paper, an algorithm based on SWR is proposed to generate a Chinese text classification adversarial example. It divides Chinese similar words into semantically similar words and appearance-similar words and designs the generation algorithms, respectively. The generation method of appearance-similar words is designed by considering the Chinese character's image, the Fivepen code, the Zhengma code, and the stroke order code at the same time. By comprehensively using semantically SWR, appearance-SWR, traditional character replacement, and word order disturbance, the adversarial examples generated by this strategy have achieved good results in deceiving both Chinese texts classifiers and human observers in the experiment and human observer test. The experimental results show that SWR has enough concealment, and the modification of the texts and modification strategies are difficult to be found. The availability of the algorithm has been proved.

There are three key tasks in the next step: to improve the transferability of the algorithm, to attempt to implement a transformer based adversarial attack, and to achieve a black box attack on the target classifier. In application, it is difficult to know the specific parameters of the target classifier and the confidence of its output results in each category. It is a popular method to use new concepts such as transformer to achieve a black box adversarial attack. Therefore, completing these three tasks can enhance the practicality of the algorithm.

References

1. Goodfellow, I., Shlens, J., Szegedy, C.: Explaining and harnessing adversarial examples. In: Proceedings 3rd International Conference on Learning Representations, ICLR 2015, San Diego, CA, USA (2015)
2. Alzantot, M., Sharma, Y., Elgohary, A., et al.: Generating natural language adversarial examples. In: Proceedings of the 2018 Conference on Empirical Methods in Natural Language Processing (2018)
3. Papernot, N., McDaniel, P., Swami, A., et al.: Crafting adversarial input sequences for recurrent neural networks. In: MILCOM 2016–2016 IEEE Military Communications Conference, pp. 49–54 (2016)

4. Samanta, S., Mehta, S.: 'Towards crafting text adversarial examples'. arXivLabs, (2017). https://doi.org/10.48550/arXiv:1707.02812.
5. Ebrahimi, J., Rao, A., Lowd, D., et al.: Hotflip: white-box adversarial examples for text classification. In: The 56th annual meeting of the association for Computational Linguistics, vol. 2, pp. 31–36 (2018)
6. Liang, B., Li, H., Su, M., et al.: Deep Text Classification Can Be Fooled. In: The 27th International Joint Conference on Artificial Intelligence, pp. 4208–4215 (2018)
7. Gao, J., Lanchantin, J., Soffa, M., et al.: Black-box generation of adversarial text sequences to evade deep learning classifiers. In: IEEE Sec. Privacy Workshops (SPW), pp. 50–56 (2018)
8. Li, J., Ji, S., Du, T., et al.: 'TextBugger: Generating adversarial text against real-world applications. In: 26th Annual Network and Distributed System Security Symposium (2019)
9. Jin, D., Jin, Z., Zhou, J., et al.: 'Is BERT Really Robust? Natural Language Attack on Text Classification and Entailment'. arXivLabs, ArXiv:1907.11932, (2019)
10. Chenglong, L.: Generating adversarial text based on genetic algorithm. In: Huazhong University of Science and Technology, Wuhan, China (2019)
11. Ren, S., Deng, Y., He, K., et al.: Generating natural language adversarial examples through probability weighted word saliency. Meet. Assoc. Comput. Linguist. **2019**, 1085–1097 (2019)
12. Yuan, F., Zhang, T., Liang, X., et al.: An improved generation method of adversarial example to deceive NLP deep learning classifiers. In: Artificial Intelligence for Communications and Networks: Second EAI International Conference (AICON), vol. 356, pp. 51–62 (2020)
13. Wang, W., Wang, R., Wang, L., et al.: Adversarial examples generation approach for tendency classification on Chinese texts. J. Software. **30**(8), 2415–2427 (2019)
14. Tong, X., Wang, L., Wang, R., et al.: A generation method of word-level adversarial samples for Chinese text classification. Netinfo Security. **20**(9), 12–16 (2020)
15. Xie, Y., Gu, Z., Zhu, B., et al.: Adversarial examples for Chinese text classification. In: IEEE Fifth International Conference on Data Science in Cyberspace (DSC), vol. 10(1109), pp. 238–245 (2020)
16. Nuo, C., Chang, G., Gao, H., et al.: WordChange: adversarial examples generation approach for Chinese text classification. IEEE Access. **8**, 79561–79572 (2020)
17. Li, L., Shao, Y., Song, D., et al.: 'Generating Adversarial Examples in Chinese Texts Using Sentence-Pieces'. arXivLabs, arXiv:2012.14769. 10.48550. 10.48550. (2020)
18. Zheng, Y., Cui, Z., Xu, Y., et al.: Generating adversarial examples for sentiment classifier of Chinese sentences. In: 6th International Symposium on System and Software Reliability (ISSSR), vol. 10(1109), pp. 27–32 (2020)
19. Zhang, Z., Liu, M., Zhang, C., et al.: Argot: generating adversarial readable Chinese texts. In: Twenty-Ninth International Joint Conference on Artificial Intelligence and Seventeenth Pac. Rim International Conference on Artificial Intelligence, vol. 10(24963), pp. 2533–2540 (2020)
20. Xu, E., Zhang, X., Wang, Y., et al.: Adversarial examples generation method for Chinese text classification. Int. J. Network Secur. **24**(4), 587–596 (2022)
21. Jin, R., Wu, C.: WordErrorSim: An adversarial examples generation method in Chinese by erroneous knowledge. In: The 5th International Conference on Compute and Data Analysis, pp. 155–161 (2021)
22. Xie, Y., Gu, Z., Zhu, B., et al.: Adversarial examples for Chinese text classification. In: 2020 IEEE Fifth International Conference on Data Science in Cyberspace (DSC), pp. 238–245 (2020)
23. Liu, M., Zhang, Z., Zhang, Y., et al.: Automatic generation of adversarial readable Chinese texts. IEEE Trans. Depend. Secure Comput. **20**(2), 1756–1770 (2022)
24. Zhang, S., Wu, H., Zhu, G., et al.: Character-level adversarial samples generation approach for Chinese text classification. J. Electr. Inform. Technol. **45**(6), 2226–2235 (2023)
25. Li, L., Shao, Y., Song, D., et al.: 'Generating adversarial examples in Chinese texts using sentence-pieces'. arXiv Preprint arXiv:2012.14769, 2020
26. Gao, H., Zhang, H., Yang, X., et al.: Generating natural adversarial examples with universal perturbations for text classification. Neurocomputing. **471**, 175–182 (2022)

27. Zheng, Y., Cui, Z., Xu, Y., et al.: Generating adversarial examples for sentiment classifier of Chinese sentences. In: 2020 6th international symposium on system and software reliability (ISSSR), pp. 27–32 (2020)
28. Kusner, M., Sun, Y., Kolkin, N., et al.: From word embeddings to document distances. Int. Conf. Machine Learn. **37**, 957–966 (2015)
29. Qi, J., Long, H., Shao, Y., et al.: Research on Chinese character similarity algorithm based on eigenvector and stroke coding. J. Chongqing Univ. Posts Telecommun. (Natural Science Edition). **31**(06), 885–891 (2019)
30. Ye, J.: The influence of Chinese word order on the reading. J. Chongqing University of Arts and Sciences (Social Sciences Edition). **33**(6), 77–81 (2014)

Overview of Consensus Mechanism in Blockchain

Yibei Li, Jingyi Yang, Yiwei Lai, and Mingzhe Liu

1 Introduction

The background of the research concerning consensus mechanism can be traced back to the inception of Bitcoin (BTC). On October 31, 2008, the release of a paper titled "Bitcoin: A Peer-to-Peer Electronic Cash System" marked the birth of Bitcoin, the first cryptocurrency [1].

As a decentralized digital currency, Bitcoin implements a consensus mechanism called Proof of Work (PoW). As Bitcoin network has progressed and transaction volume has escalated, certain drawbacks associated with PoW have gradually manifested, including issues like excessive energy consumption [2] and network congestion [3].

With the development and popularization of blockchain technology, more and more consensus mechanisms have been proposed and applied to address the drawbacks of PoW [4]. The goal is to resolve the limitations of PoW and provide more secure, efficient, and environmentally friendly consensus solutions. In addition, as the application scenarios of blockchain technology continue to expand and deepen [5], researchers have also started exploring consensus mechanisms suitable for different scenarios [6]. For example, consensus mechanisms with high throughput and low transaction costs suitable for enterprise-level scenarios [7], or those ensuring device security and trustworthiness suitable for IoT scenarios [8].

Another major factor influencing the research on blockchain consensus mechanisms is the policies and regulations of different countries. As the application of blockchain technology becomes more widespread, more and more countries and regions are beginning to formulate relevant policies [9], which impose higher

Y. Li (✉) · J. Yang · Y. Lai · M. Liu
Department of Computer and Cyber Security, Chengdu University of Technology, Chengdu, China

X. Jia et al. (eds.), *2nd International Conference on Big Data, IoT, and Cloud Computing*, EAI/Springer Innovations in Communication and Computing,
https://doi.org/10.1007/978-3-031-93570-1_23

requirements on the research and application of blockchain consensus mechanisms. For instance, some countries and regions have implemented restrictive measures on cryptocurrencies that adopt PoW, while encouraging and supporting the development and application of other more environmentally friendly and efficient consensus mechanisms.

With the continuous development of blockchain technology [10], consensus mechanisms will continue to be improved to meet the demands of broader scenarios. In the future, with the widespread adoption of blockchain technology, consensus mechanisms will become a significant research direction in the field of blockchain. Continuous innovation and development will provide more possibilities for blockchain technology, and they will also be subject to higher requirements and attention from policies and laws.

2 Classic Consensus Mechanisms

2.1 Proof of Work

Proof of Work (PoW) was the earliest consensus mechanism in blockchain [1]. In PoW, participants (referred to as miners) must perform specific calculations [11] to demonstrate the legitimacy and validity of the blocks they have packaged, which contain transaction records within the system. These calculation tasks are intentionally designed to be complex, demanding substantial computing power and energy resources [12]. This characteristic effectively regulates the frequency of transaction verification, preventing the blockchain network from potential attacks by malicious nodes.

Now, let's take the Bitcoin system as an example to provide a detailed explanation of the principle of PoW. As is well-known, a block within Bitcoin system comprises two main components: block header and block body [13]. The block header consists of six fields: version, timestamp, target, nonce, previous block hash, and Merkle root. Miners in Bitcoin network bundle around a thousand recently received unconfirmed transactions into blocks and compete through mining to earn the right to record these blocks and receive block rewards. During the mining process, a miner initially set the nonce value in the block header to 0 and utilize SHA-256 to compute the hash value H(Header) for the entire block header. If the calculated hash value H(Header) is not greater than the mining difficulty target (i.e., satisfying the condition H(Header) $\leq$ target) [14], the mining process is successful, and the miner gains the right to record the block on the blockchain. In case the initial nonce value does not meet the conditions for successful mining, the miner increments the nonce value by 1 and repeats the SHA-256 hash calculation for the block header. This process continues until the miner discovers a nonce value that satisfies the conditions for successful mining or finds that other miners have already added the new block to the blockchain. In the latter scenario, the miner has to abandon the current competition and shifts focus to compete for the right to record the next block on the blockchain. The difficulty of mining increases as the target value decreases, making it more challenging to find a valid nonce [15].

The hash function SHA-256 is "puzzle-friendly" [16], meaning that miners can only adjust the nonce value through an exhaustive search to find a hash that meets the required conditions. It is not possible for miners to predict or set the nonce value in advance, ensuring a fair and competitive mining process. As a result, the probability of a miner successfully obtaining the right to record a block is directly proportional to the computing resources they invest. In other words, the more computational power a miner dedicates to the mining process, the higher their chances of winning the right to add a block to the blockchain. Conversely, if a miner has limited computing resources, their probability of winning the bookkeeping right decreases accordingly. This implies that to participate as a block-producing node, miners must invest a certain amount of computing resources [17].

The inclusion of the Merkle root hash value in block header guarantees the immutability of the content (transaction information) within block body. Consequently, calculating the hash value of block header is sufficient to ensure the overall integrity and correctness of the entire block content, preventing any possibility of tampering [18].

The introduction of PoW successfully distributes the accounting rights of the cryptocurrency system fairly to all nodes in the network, achieving decentralization. It uses block rewards and transaction fees as economic incentives, stimulating the enthusiasm of all nodes. In a blockchain network that employs PoW, malicious attacking nodes cannot successfully launch an attack if they do not control more than 51% of the network's computing power [19]; notably, the large scale of the Bitcoin network further prevents such possibility. Therefore, PoW has good security.

However, PoW also faces a series of problems, such as concentration of computing power and excessive energy consumption. Wealthy individuals with economic means can buy more high-performance mining equipment and bear enormous electricity costs, allowing them to monopolize the computing power of the entire network and deprive ordinary nodes of the opportunity to obtain accounting rights, leading to a centralization tendency in the system [20]. Furthermore, the significant energy consumption during the PoW mining process has been controversial since the emergence of Bitcoin. As the computing power in networks increases exponentially, the "target" value becomes smaller, requiring more and more electricity. This gives rise to increasingly severe environmental concerns [21].

2.2 Proof of Stake

In 2011, in response to the issue of excessive energy consumption associated with PoW, a new consensus mechanism called Proof of Stake (PoS) was proposed on the Bitcointalk forum. Unlike PoW, where competition for computing power input is paramount, PoS operates by randomly selecting a node and granting it the right of bookkeeping for the next block, and the selected node is called "validator" in PoS. The probability of a node being chosen as a validator and its "stakes"—represented by the holdings of specific assets—are directly proportional; for instance, in cryptocurrency systems utilizing PoS, validators' stake refers to the amount of cryptocurrency held by them.

In 2012, a cryptocurrency called Peercoin (PPC) was introduced [22]. Peercoin pioneered the concept of "coinage" and adopted coinage-based PoS as its consensus mechanism. Cryptocurrency systems that utilize PoS often generate a certain amount of initial currency through PoW during the early stage. Subsequently, when PoS is activated, what nodes need to invest in competing for bookkeeping rights shift from computing power to coinage. In Peercoin system, the coinage obtained by a node during a certain period of time is the accumulation of currency (stake) held by the node over the period, calculated by the formula $C = s \cdot T$ (C refers to the coinage obtained by the node; additionally, s refers to the amount of currency, and T refers to how many days the period continuing) [22]. When vying for the bookkeeping right of the next block, a node invests some coinage and then performs SHA-256 hash operation on the block header to obtain the hash value H(Header). If the condition

$$H\,(\text{Header}) \leq \text{target} \times \text{coinage}$$

is met, the block will be successfully generated [23]. It becomes evident that the more coinage a node invests, the higher the probability of obtaining the bookkeeping right. To ensure fairness in the competition for bookkeeping rights, every time a block is successfully generated, the coinage invested by the validator will be cleared.

Compared with Bitcoin system that uses PoW, Peercoin system that adopts PoS eliminates the "nonce" field in block header. This change ensures that the hash value H(Header) of block header only changes passively with the timestamp counting in seconds. As a result, it greatly alleviates the problem of computational competition and reduces the waste of electrical resources [24]. Due to not requiring expanded computational resources, PoS has better scalability compared to PoW, and could be better applied in public chains [25].

However, PoS also exhibits the Matthew effect, where wealthy validators who hold a large amount of cryptocurrency can easily accumulate a significant coinage within a short period of time. This allows them to gain an absolute advantage in the competition for accounting rights, monopolize block rewards, and go against the original intention of decentralized blockchain. Furthermore, it is also a security vulnerability in a PoS system that nodes holding over 50% of the entire network's cryptocurrencies can launch attacks on the main chain [26].

2.3 Delegated Proof of Stake

In April 2014, Dan Larimer, the chief developer of a open-source distributed trading system called Bitshares, made improvements based on PoS and introduced a new consensus mechanism called Delegated Proof of Stake (DPoS) [27]. In blockchain systems using DPoS, committees are formed by electing several nodes (representatives) through voting by all nodes. The committee members collect transactions, package and verify blocks, and receive block rewards and transaction fees.

In DPoS, the voting weight of a node is directly proportional to the amount of its stake (such as cryptocurrency). At the end of each voting round, the N nodes with the highest weighted votes are elected as representatives (the value of N is determined by the initiator of the blockchain project) [28], and all nodes in the network have the right to vote and compete to become a representative. During the operation of DPoS, each representative is allocated a time slot ΔT, and they can only acquire the right to account for new blocks within their allocated time slot. After a duration of $N \cdot \Delta T$, which is the total time allocated to all representatives, the system starts the next round of voting to re-elect N representatives, and this process continues indefinitely.

Compared to PoS, DPoS concentrates the generation of new blocks within the committee [29], increasing block efficiency and saving a significant amount of computing resources. Moreover, since networks utilizing DPoS are easily scalable, DPoS applicable to public chains [30]. However, similar to PoS, DPoS also exhibits a phenomenon where nodes with a large stake tend to have more advantages. In a cryptocurrency system using DPoS, nodes holding a substantial amount of currency may collaborate for their common interests, vote for each other, and exploit the advantage of stake weight to monopolize voting power, control committee members, and dominate block rewards, leading to an increasing concentration of wealth in a few individuals. Furthermore, nodes with a massive stake can vote for themselves and use their weight advantage to secure a decisive number of votes.

2.4 Practical Byzantine Fault Tolerance

In a distributed network, Byzantine Generals Problem pertains to the challenge of reaching a consensus among honest nodes regarding data, under the premise that certain nodes are known to have crashed or are malicious [31]. In 1999, Miguel Castro and Barbara Liskov proposed a consensus mechanism named Practical Byzantine Fault Tolerance (PBFT) as a solution to Byzantine Generals Problem in distributed systems [32]. In a distributed network system, PBFT ensures that when the number of faulty or malicious nodes (referred to as Byzantine nodes) is less than one-third of the total number of nodes in the network, denoted as $n \geq 3f + 1$, the network achieves consensus on messages [32].

Consensus Process

In a C-S (client-server) network model with a distributed architecture on the server side using PBFT, there are two types of roles: the primary node and the replica nodes. Each round of the consensus process selects a primary node from all the nodes in the network, with the rest being replicas. The consensus process consists of five stages: Request, Pre-prepare, Prepare, Commit, and Reply. The execution of each stage is as follows:

(I) Request: The client sends a request message

$$m =< \text{REQUEST}, o, t, c >$$

Along with a signature of the request message encrypted using the client's private key, to the primary node. The REQUEST signifies the current protocol stage of the message, "o" represents the specific operation requested by the client, "t" denotes the timestamp appended when the client sends the request, and "c" serves as the identifier for the client.

(II) Pre-prepare: The primary node receives a request message "m" from the client. If there are too many requests from clients at the same time, the primary node will cache the request messages for later processing. The primary node then verifies the signature of the request message by decrypting it using the client's public key. If the signature verification is successful, the primary node calculates the pre-prepare message

$$<< \text{PREPREPARE}, d, s, v >, m >$$

And signs <PRE_PREPARE, d, s, v>. It then broadcasts the aforementioned message to the replica nodes. In the message, d represents H(m) (H refers to a hash function), the hash value of the request message "m", "s" is a message sequence number used to order the client's requests, and "v" is a view number used to track changes in the primary node (the system assigns a numerical identifier to each of the R nodes as 0, 1, 2, ..., R-1. The current primary node's identifier is v mod R. Whenever the primary node changes, the view number increments by 1).

(III) Prepare: Upon receiving the pre-prepare message from the primary node, each replica node performs the following verifications:

1. Validate the signature of the message received from the primary node.
2. Ensure that d = H(m).
3. Ensure that the current view number matches the view number field "v" in the pre-prepare message.
4. Check if they have never received a pre-prepare message with the same view number and message sequence number fields as the pre-prepare message, but with a different hash value field. Since it is impossible for the primary node to send two pre-prepare messages with the same view number and message sequence number but different request messages and hash values.

5. The value of message sequence number s is between h and H (h and H are known as the low-water mark and high-water mark of the message sequence number).

If all the above conditions are verified, send the prepare message

$$< \text{PREPARE}, d, s, v, i >$$

And its signature to all nodes, where the hash value, message sequence number, and view number fields are the same as those in the pre-prepare message, and "i" is the number of the replica node sending the prepare message.

(IV) Commit: The primary node and replica nodes perform the following validations for each received prepare message:

1. Verify the signature of the prepare message.
2. Ensure that d = H(m).
3. Ensure the view number field is consistent with the current view number.
4. Ensure the value of the message sequence number is between h and H.

The pre-prepare message and all validated prepared messages are written into the message log for use in the view change protocol (which will be explained later). Assuming the total number of nodes in the network, R, satisfies $R = 3f + 1$, where f is the maximum number of tolerated Byzantine nodes by the system. If a node receives a number of valid prepare messages greater than or equal to $2f + 1$ (including the prepare messages it sent itself, meaning that a normal node only needs to receive 2f valid prepare messages), it will broadcast a commit message

$$< \text{COMMIT}, d, s, v, i >$$

and its signature to all nodes. The hash value, message sequence number, view number, and node number fields in the commit message are the same as those in the prepare messages sent in the previous phase.

(V) Reply:The primary node and replica nodes perform the following validations for each received commit message:

1. Verify the signature of the commit message.
2. Ensure that d = H(m).
3. Ensure the view number field is consistent with the current view number.
4. Ensure the value of the message sequence number is between h and H.

All validated commit messages are written into the log. If a node receives no less than $2f + 1$ valid commit messages (including the commit message it sent itself), it indicates that consensus has been reached by more than two-thirds of the network's nodes. The node then proceeds to execute the requested operation from the client and sends a reply message

$$< \text{REPLY}, t, c, v, i, r >$$

back to the client, where r represents the result of the requested operation. If the client receives f + 1 reply messages, it signifies the successful completion of one round of consensus. In a system where the number of Byzantine nodes is less than one-third of the total network nodes, there are at most f Byzantine nodes. Therefore, when the client receives f + 1 reply messages, it means that at least one honest node is included. In the previous phases, honest nodes have received no less than 2f + 1 valid commit messages, so their reply messages can represent consensus from more than two-thirds of the honest nodes in the network. Otherwise, the client needs to resend the request to the primary node.

Advantages and Disadvantages

Compared to PoW, PBFT is not a high-energy consumption consensus mechanism [33], making it more environmentally friendly and energy-efficient. Under the premise of a relatively small number of nodes, PBFT has high efficiency, allowing the network nodes to reach consensus quickly. Additionally, networks using PBFT are not constrained by nodes with low throughput capabilities. Due to the adoption of mechanisms such as multi-round voting and digital signature, PBFT also has good security.

However, PBFT also has several drawbacks. In a round of PBFT consensus, each normal node needs to communicate with all other normal nodes via P2P, so as the number of nodes n increases, the communication time complexity of PBFT is $O(n^2)$ [34]. Although the polynomial time complexity makes Byzantine fault-tolerant algorithm feasible in practical systems, the quadratic complexity is still relatively high [35], which hampers the scalability of distributed network systems using PBFT. Particularly, when the number of system nodes reaches around 100, the performance degradation becomes significant [36], which is why PBFT cannot be applied to public blockchains and is only suitable for consortium or private blockchains. In a network environment with unstable communication quality, PBFT, which requires frequent P2P consensus synchronization between each pair of nodes, also suffers from high latency disadvantages. Compared to consensus mechanisms like PoW and PoS, PBFT has lower fault tolerance. The former achieves 50% fault tolerance (to guard against 51% computing power attack or 51% token attack), while the latter only achieves one-third fault tolerance, which means that if the proportion of malicious nodes reaches 34%, they can launch an attack on the system.

3 Other Consensus Mechanisms

In addition to the aforementioned several classic consensus mechanisms, various other consensus solutions have been proposed with the development of blockchain technology. The aim is to provide good solutions for ensuring the operational efficiency, security, fairness, and scalability of blockchain systems.

3.1 Proof of Elapsed Time

Proof of Elapsed Time (PoET), proposed by Intel in 2016, is a consensus mechanism based on hardware randomness [37]. PoET simulates the time consumed by PoW mining by using random hardware delay to select validators in a blockchain network. The consensus process of PoET is as follows:

(I) Generation of waiting time [38]: In order to elect the validators for the current round, which are responsible for generating new blocks, each node generates a random waiting time using a secure encryption algorithm. The waiting time is calculated within Trusted Execution Environment (TEE) [39], ensuring that it remains hidden from other nodes and cannot be tampered with.

(II) Waiting: Nodes start timers and wait for the specified duration of the waiting time they generated.

(III) Validator establishment: The node with the shortest waiting time, which finishes the waiting phase first, is selected as the validator for this round.

(IV) Block generation: The node selected as the validator gains the right to record, generates the next block in the blockchain, and broadcasts it to the network. Other nodes in the blockchain verify the validity of the digital signatures and transaction records when they receive the block broadcasted by the validator. If the block is verified as valid by the entire network, it is successfully added to the blockchain.

The key to the PoET consensus process lies in the selection of waiting time, which relies on the hardware's randomness to ensure equal opportunities for every participant to become a validator. Compared to other consensus algorithms, PoET has advantages such as low energy consumption and good scalability [40]. Therefore, PoET does not require a large amount of computing resources and energy consumption, making it suitable for large-scale networks like public chains [41]. PoET does not face performance bottlenecks as the network scales. Additionally, thanks to the application of hardware randomness, PoET is also efficient. However, PoET does have some potential issues. Due to its reliance on specific hardware, such as Intel SGX, PoET faces centralization risks and relies heavily on users' trust in a single authority institution.

3.2 Proof of Authority

Proof of Authority (PoA) was proposed by Gavin Wood, co-founder of Ethereum, as an alternative to PoW and PoS. It is a consensus mechanism based on authoritative nodes [42]. The nodes participating in the PoA consensus process are called validators. Unlike PoS, validators in PoA stake their identity instead of tokens and need to establish authority through a mandatory authentication process [43]. Validators publicly disclose all proof documents and demonstrate their consistent ability to contribute to consensus in order to maintain the privilege of participating in the block validation process. Generally, in the PoA consensus process, a group of validators is elected by the administrators or participants of the blockchain network based on factors such as identity, reputation, and expertise. Validators are responsible for verifying transactions in the network and packaging them into new blocks. Each validator takes turns to have the right to record, ensuring the validity and legitimacy of transactions with their highly trusted identity and authority. Therefore, the confirmation stage of the PoA consensus process is relatively simple, where the signature or confirmation of validators is sufficient to gain consensus recognition from the entire network, significantly reducing the time delay in the consensus process.

PoA is characterized by its small and trusted consensus group, sacrificing decentralization for system security and performance [44]. Compared to classical consensus mechanisms like PoW, PoA has very low requirements for system computational power, and high throughput since only a few validators participating in very short processes. However, in a network using PoA, due to the lack of oversight from regular nodes towards validators, theoretically, validators can easily collude and launch attacks on the network [45]. Because of this reason, validators have significant control over the entire network, leading to a strong centralization tendency [44], contradicting the original intention of decentralization in blockchain. Additionally, systems employing PoA have weak resistance against external attacks. Once validator nodes are compromised, the system is directly threatened. Due to its high dependency on validators' authority, PoA is almost only suitable for scenarios within organizations or institutions, such as alliance chains. Its highly centralized property also makes it unsuitable for application in public chains.

3.3 Proof of Capacity

Proof of Capacity (PoC) was proposed in 2014. Unlike traditional consensus mechanisms, PoC selects block creators based on storage capacity [46]. Nodes participating in PoC consensus process need to allocate a certain amount of storage space in advance, which can come from storage media such as hard drives. During the consensus process, nodes prove that they have a sufficiently large storage capacity to earn the right to add blocks to the blockchain. The consensus process of PoC is as follows:

(I) Storage space allocation: Nodes participating in the consensus allocate a certain amount of storage space to the PoC algorithm and initialize it. The storage space is divided into a series of equally sized data blocks, also known as "sectors."

(II) Pre-calculation: Nodes use a random number generator to fill each sector with a corresponding set of seed values, and then calculate the hash value for each sector's data (using a hash function, usually Shabal).

(III) Iterative calculation: The previous round's calculated hash values are used as inputs to the hash function, and a new set of hash values is calculated. This process is iterated repeatedly until a hash value is found that meets certain conditions (such as being less than a specific value or having a specific number of leading zeros), indicating a successful mining attempt.

Compared to PoW, PoC mainly relies on storage capacity rather than computational power. Participants only need to allocate a certain amount of storage space in advance, without the need for extensive computational tasks. Therefore, the energy consumption is significantly lower than that of PoW [47]. Since the cost of participating in mining is mainly the allocation of storage capacity, attackers would need to control a large amount of physical storage devices throughout the network in order to dominate, which is obviously more difficult than controlling a large amount of computational power. Therefore, a network using PoC is more resistant to 51% attacks and has higher security than a network using PoW. Additionally, if there is a need to increase the network size, it is only necessary to add more storage devices, without the need for complex expansion of computational resources. Therefore, a network using PoC also has good scalability. Due to its combination of security and scalability, PoC could be applied to large-scale scenarios such as public chains.

However, PoC also has shortcomings such as high initial costs and maintenance costs for nodes. Furthermore, PoC does not solve the problem of centralization risk, meaning that wealthy participants can still monopolize more accounting rights by purchasing a large number of storage devices, resulting in the concentration of wealth throughout the network.

3.4 Combinations of Consensus Mechanisms

PoW/PoS

Proof of Work (PoW) and Proof of Stake (PoS) can be combined in various ways, depending on the requirements of different blockchain projects, with the aim of improving performance and security [48]. One common combination involves initially using PoW in systems' early stages [49] to ensure decentralization and provide initial security. Later, a transition is made to PoS, which offers higher efficiency and scalability. Another common combination of PoW and PoS is deep integration, where both mechanisms are used consecutively in each consensus round. During the block generation phase, PoW is used, where nodes acting as miners generate new blocks through mining. In the phase of block confirmation, PoS is utilized. Nodes acting as validators verify new blocks. The system first elects a group of validators based on token holdings (the specific election algorithm varies among different systems, with nodes holding more tokens having a higher probability of being chosen as validators). Then, these validators vote "yes" or "no" on the validity of the new block. Validators who hold more tokens have greater voting power. If a certain proportion of weighted votes are in favor, the new block is added to the blockchain. Since different roles serve as miners and validators in the same consensus round, a balance is achieved between computational power holders and token holders, thereby enhancing the fairness and security of the blockchain.

Due to its higher level of decentralization compared to standalone PoW and PoS, the deep integration is suitable for large-scale scenarios such as public chains that require a high degree of decentralization.

Tendermint

Tendermint—a consensus mechanism, proposed by Jae Kwon in 2014, is a combination of PoS and PBFT [7]. It is a Byzantine fault-tolerant algorithm based on proposal and voting, aiming to address the issues of high-energy consumption, low throughput, and difficulty in quantifying security in PoW systems. The system first elects a subset of nodes as validators to form a committee (the election can be based on stake, i.e., the amount of cryptocurrency held by the nodes). As the consensus process progresses through rounds, different validators from the committee would be selected by the system as "proposer"—responsible for generating new blocks [50].

After the proposer generates a block and broadcasts it to the entire network, the validators verify its validity. If the verification passes, it proceeds to the pre-voting phase. If the verification fails or times out, the block generated by the proposer is discarded, and the system generates an empty block (without any transaction information) to proceed to the pre-voting phase based on this empty block. The first vote (pre-vote) is participated by the committee. If more than two-thirds of the validators vote in favor, it directly enters the pre-commit phase. Otherwise, after the system generates an empty block, it enters the pre-commit phase based on the empty block. The pre-commit phase is the second vote (formal voting), still involving the committee. If more than two-thirds of the validators vote in favor, the block is successfully added to the blockchain. Otherwise, the block is discarded, and the current round of consensus ends.

Due to the fact that both voting processes during the consensus are only participated by the committee members, not all nodes, Tendermint has high efficiency and a large system throughput [50]. However, due to the incorporation of PBFT, Tendermint has low fault tolerance, allowing a maximum of only 33% of Byzantine nodes of the number of committee members.

Tendermint can achieve high throughput, and due to its low resource overhead in the consensus process, transaction costs could also be reduced. Therefore, Tendermint is suitable for enterprise-level large-scale scenarios.

4 Outlook

This paper introduces various consensus algorithms and analyzes their advantages and disadvantages in terms of decentralization, scalability, efficiency, cost, energy consumption, security, and so on. Based on such analysis, it deduces the suitable scenarios and applicable scales for each algorithm. According to our summary, currently, the research of consensus mechanisms still faces several challenges:

1. Scalability and throughput: As the number of users and transactions grows significantly, many consensus mechanisms face bottlenecks in terms of throughput and scalability [51]. Future research and innovation will aim to overcome these challenges by proposing highly scalable and high-throughput mechanisms [52] that can support large-scale business scenarios to provide global services.
2. Environmental sustainability: Traditional consensus mechanisms, such as PoW, consume a significant amount of energy, posing pressure on the environment. Driven by relevant policies worldwide, future consensus mechanism research will focus on improving energy efficiency [53] and even integrating technologies like renewable energy. This aims to reduce the reliance on high-energy consumption in blockchain applications and mitigate the negative environmental impact.
3. Security and decentralization: Aligned with the decentralization concept of blockchain technology, future blockchain research will prioritize the design and implementation of decentralized governance models [54]. This includes developing community decision-making mechanisms and involving more participants to enhance the democratic and transparent nature of the consensus process. These efforts aim to promote community autonomy and sustainable development. Therefore, ensuring network security and decentralization is one of the central goals of consensus mechanism research. Future research will continue to prioritize system security, strengthen resilience against various network attacks [55], and explore better strategies to preserve the decentralized nature of the consensus process, preventing power concentration and potential manipulation.
4. Interoperability among blockchains: As blockchain platforms are increasingly applied in various industries, the interaction between different blockchains has become an important issue. Future consensus mechanism research will heavily focus on promoting interoperability between blockchain platforms, enabling secure and reliable data transfer and asset transfer between different blockchains [56]. This will facilitate seamless integration of two or more blockchains.

These challenges also serve as trends that guide future blockchain research and provide a continuous source of motivation for the development of blockchain technology. In the future, blockchain technology could evolve towards higher performance, better scalability, lower energy consumption, improved environmental sustainability, enhanced security, larger-scale decentralized governance, and increased cross-chain interoperability.

Additionally, it is worth noting that the combination of blockchain and IoT has emerged as a significant research focus within the field of blockchain in recent years [57]. However, previous consensus mechanisms have primarily concentrated on environments with high-energy consumption and low throughput. Consequently, there is a pressing need to develop lightweight consensus mechanisms suitable for resource-constrained IoT devices. Another critical aspect of consensus mechanism design is to reduce the vulnerability to attacks by malicious nodes during the consensus process and enhance overall system reliability through comprehensive review and evaluation schemes. Machine learning applications in consensus mechanisms have also emerged as a prominent research direction. Machine learning algorithms can be leveraged for reviewing and evaluating behaviors, detecting malicious activities, and eliminating malicious nodes to bolster system security. Additionally, they can potentially replace energy-intensive PoW mining, effectively reducing energy consumption in the process.

References

1. Nakamoto, S. (2008). Bitcoin: a peer-to-peer electronic cash system, Decentralized business review 21260
2. Zhang, R., Chan, W.K.V.: Evaluation of energy consumption in block-chains with proof of work and proof of stake. J. Phys. Conf. Ser. **1584**(1), 012023 (2020)
3. Kiffer, L., et al.: Security of nakamoto consensus under congestion. Cryptology ePrint Archive (2023)
4. Hinzen, F.J., John, K., Saleh, F.: Proof-of-work's limited adoption problem. NYU Stern School of Business (2019)
5. Guo, Y., Liang, C.: Blockchain application and outlook in the banking industry. Financ. Innov. **2**, 1–12 (2016)
6. de Oliveira, M.T., et al.: Blockchain reputation-based consensus: a scalable and resilient mechanism for distributed mistrusting applications. Comput. Netw. **179**, 107367 (2020)
7. Buchman, E. (2016). Tendermint: Byzantine fault tolerance in the age of blockchains (Doctoral dissertation, University of Guelph)
8. Rawat, A., Daza, V., Signorini, M.: Offline scaling of IoT devices in IOTA blockchain. Sensors. **22**(4), 1411 (2022)
9. Novak, M.: Crypto-friendliness: understanding blockchain public policy. J. Entrepreneurship Public Policy. **9**(2), 165–184 (2020)
10. Johannes, S., et al. "Recent developments in blockchain technology and their impact on energy consumption." arXiv preprint arXiv:2102.07886 (2021)
11. Sobti, R., Geetha, G.: Cryptographic hash functions: a review. Int. J. Comput. Sci. Iss. (IJCSI). **9**(2), 461 (2012)
12. Gilbert, H., Handschuh, H.: Security analysis of SHA-256 and sisters. In: International Workshop on Selected Areas in Cryptography. Springer, Berlin, Heidelberg (2003)
13. Vivekanadam, B.: Analysis of recent trend and applications in block chain technology. J. ISMAC. **2**(04), 200–206 (2020)
14. Westerkamp, M., Eberhardt, J.: Zkrelay: facilitating sidechains using zksnark-based chain-relays. In: 2020 IEEE European Symposium on Security and Privacy Workshops (EuroS&PW). IEEE (2020)
15. Zhang, X., et al.: An analysis of blockchain-based bitcoin mining difficulty: techniques and principles. In: 2018 Chinese Automation Congress (CAC). IEEE (2018)
16. Wang, M., Duan, M., Zhu, J.: Research on the security criteria of hash functions in the blockchain. In: Proceedings of the 2nd ACM Workshop on Blockchains, Cryptocurrencies, and Contracts (2018)
17. Andrychowicz, M., Dziembowski, S.: Pow-based distributed cryptography with no trusted setup. In: Advances in Cryptology–CRYPTO 2015: 35th Annual Cryptology Conference, Santa Barbara, CA, USA, August 16–20, 2015, Proceedings, Part II 35. Springer, Berlin (2015)
18. Chenchev, I.: Speedup of Merkle-Root Hash Value Computation Using Groups. In: Future of Information and Communication Conference. Springer Nature Switzerland, Cham (2023)
19. Aponte-Novoa, F.A., et al.: The 51% attack on blockchains: a mining behavior study. IEEE Access. **9**, 140549–140564 (2021)

20. Beikverdi, A., Song, J.S.: Trend of centralization in bitcoin's distributed network. In: 2015 IEEE/ACIS 16th International Conference on Software Engineering, Artificial Intelligence, Networking and Parallel/Distributed Computing (SNPD). IEEE (2015)
21. Rafiqul, I.M., et al.: A comprehensive analysis of blockchain-based cryptocurrency mining impact on energy consumption. Int. J. Adv. Comput. Sci. Appl. **13**, 4 (2022)
22. King, S., Nadal, S.: "Ppcoin: Peer-to-peer crypto-currency with proof-of-stake". self-published paper, August 19.1 (2012)
23. Zhang, S., Lee, J.-H.: Analysis of the main consensus protocols of blockchain. ICT Express. **6**(2), 93–97 (2020)
24. Sriman, B., Ganesh Kumar, S., Shamili, P.: Blockchain technology: Consensus protocol proof of work and proof of stake. In: Intelligent Computing and Applications: Proceedings of ICICA 2019, p. 2021. Springer, Singapore
25. Yang, J., et al.: A proof-of-stake public blockchain based pricing scheme for peer-to-peer energy trading. Appl. Energy. **298**, 117154 (2021)
26. Lee, S., Kim, S.: Proof-of-stake at stake: predatory, destructive attack on PoS cryptocurrencies. In: Proceedings of the 3rd Workshop on Cryptocurrencies and Blockchains for Distributed Systems (2020)
27. Luo, Y., et al.: A new election algorithm for DPoS consensus mechanism in blockchain. In: 2018 7th International Conference on Digital Home (ICDH). IEEE (2018)
28. Liu, Y., Guangxia, X.: Fixed degree of decentralization DPoS consensus mechanism in blockchain based on adjacency vote and the average fuzziness of vague value. Comput. Netw. **199**, 108432 (2021)
29. Chaudhury, A.: "RepuStake: a delegated proof of stake protocol that stores the past behavior of witnesses." Available at SSRN 3998016 (2021)
30. Gao, W., et al.: Improved byzantine fault-tolerant algorithm based on Alliance chain. Wireless Commun. Mob. Comput. **2021**, 1–10 (2021)
31. Lamport, L., Shostak, R., Pease, M.: The byzantine generals problem. In: Concurrency: the works of leslie lamport, pp. 203–226 (2019)
32. Castro, M., Liskov, B.: Practical byzantine fault tolerance. OsDI. **99**, 1999 (1999)
33. Li, W., et al.: A scalable multi-layer PBFT consensus for blockchain. IEEE Trans Parallel Distrib Syst. **32**(5), 1146–1160 (2020)
34. Jiang, Y., Lian, Z.: High performance and scalable byzantine fault tolerance. In: 2019 IEEE 3rd Information Technology, Networking, Electronic and Automation Control Conference (ITNEC). IEEE (2019)
35. Pan, Y., et al.: Hybrid consensus algorithm based on hierarchy authority. In: IEEE International Conference on Electronic Technology, Communication and Information (ICETCI), p. 2021. IEEE (2021)
36. Yang, J., et al.: Improved fault-tolerant consensus based on the PBFT algorithm. IEEE Access. **10**, 30274–30283 (2022)
37. Bowman, M., et al.: On elapsed time consensus protocols. In: Progress in Cryptology–INDOCRYPT 2021: 22nd International Conference on Cryptology in India, Jaipur, India, December 12–15, 2021, Proceedings 22. Springer International Publishing (2021)
38. Aggarwal, Shubhani, and Neeraj Kumar. "Cryptographic consensus mechanisms." Advances in Computers. 121. Elsevier, 2021. 211–226
39. Sabt, M., Achemlal, M., Bouabdallah, A.: Trusted execution environment: what it is, and what it is not. In: 2015 IEEE Trustcom/BigDataSE/Ispa, vol. 1. IEEE (2015)
40. Corso, A.: Performance Analysis of Proof-of-Elapsed-Time (Poet) Consensus in the Sawtooth Blockchain Framework. Diss. University of Oregon (2019)
41. Zhao, W., Yang, S., Luo, X.: On consensus in public blockchains. In: Proceedings of the 2019 International Conference on Blockchain Technology (2019)
42. Joshi, S.: "Feasibility of proof of authority as a consensus protocol model." arXiv preprint arXiv:2109.02480 (2021)
43. Anushree, A.A., Saxena, A.: Two hop blockchain model: resonating between proof of work (PoW) and proof of authority (PoA). International Journal of Information Systems & Management Science. **1**, 1 (2018)

44. Fahim, S., Rahman, S.K., Mahmood, S.: Blockchain: a comparative study of consensus algorithms PoW, PoS, PoA, PoV. Int. J. Math. Sci. Comput. **3**, 46–57 (2023)
45. Wang, Q., et al.: Exploring unfairness on proof of authority: Order manipulation attacks and remedies. In: Proceedings of the 2022 ACM on Asia Conference on Computer and Communications Security, p. 2022
46. Jiang, S., Jie, W.: A game-theoretic approach to storage offloading in PoC-based mobile blockchain mining. In: Proceedings of the Twenty-First International Symposium on Theory, Algorithmic Foundations, and Protocol Design for Mobile Networks and Mobile Computing (2020)
47. Tanwar, S.: Distributed consensus for Permissionless environment. In: Blockchain Technology: from Theory to Practice, pp. 153–189. Springer Nature Singapore, Singapore (2022)
48. Gupta, K.D., et al.: A hybrid POW-POS implementation against 51 percent attack in cryptocurrency system. In: 2019 IEEE International Conference on Cloud Computing Technology and Science (CloudCom). IEEE (2019)
49. dos Santos Renato P., and Melanie Swan. "PoW, PoS, & Hybrid protocols: A Matter of Complexity?." arXiv preprint arXiv:1805.08674 (2018)
50. Kwon, J.: Tendermint: Consensus without mining. Draft v. 0.6, fall. **1.11**, 1–11 (2014)
51. Sohrabi, N., Tari, Z.: On the scalability of blockchain systems. In: 2020 IEEE International Conference on Cloud Engineering (IC2E). IEEE (2020)
52. Xu, Y., et al.: E-commerce blockchain consensus mechanism for supporting high-throughput and real-time transaction. In: Collaborate Computing: Networking, Applications and Worksharing: 12th International Conference, CollaborateCom 2016, Beijing, China, November 10–11, 2016, Proceedings 12. Springer International Publishing (2017)
53. Bada, A.O., et al.: Towards a green blockchain: A review of consensus mechanisms and their energy consumption. In: 2021 17th International Conference on Distributed Computing in Sensor Systems (DCOSS). IEEE (2021)
54. Chen, Y., Bellavitis, C.: Blockchain disruption and decentralized finance: The rise of decentralized business models. J. Bus. Ventur. Insights. **13**, e00151 (2020)
55. Sayeed, S., Marco-Gisbert, H.: Assessing blockchain consensus and security mechanisms against the 51% attack. Appl. Sci. **9**(9), 1788 (2019)
56. Pillai, B., Biswas, K., Muthukkumarasamy, V.: Cross-chain interoperability among blockchain-based systems using transactions. The Knowl. Eng. Rev. **35**, e23 (2020)
57. Li, C., et al.: Lightweight blockchain consensus mechanism and storage optimization for resource-constrained IoT devices. Inf. Process. & Manag. **58**(4), 102602 (2021)

Design of a Restaurant Review Analysis System

Wenkai Wang, Jiachao Niu, Xufeng Ling, and Ju Feng

1 Introduction

In today's society, consumers are paying more and more attention to word-of-mouth and experience when choosing restaurants. In the Internet era, mobile users' account for the majority of food and beverage consumption, and most of the existing studies use semantic analysis systems to analyze the sentiment of review content, or introduce hierarchical clustering into online reviews, and extract other useful factors that affect the score based on the collected data [1–3].

Through the restaurant review analysis system, restaurants can understand the needs and expectations of consumers, find and improve problems in service in time, and enhance the competitiveness of restaurants.

The restaurant review analysis system refers to the system that collects, sorts, analyzes, and displays the reviews about restaurants published by users on the Internet platform by using natural language processing, data mining, and other technical means. The system can capture users' evaluations of restaurant dishes, services, environment, and other aspects in real time, provide decision-making support for restaurant management, and help restaurants improve service and customer satisfaction.

The restaurant review analysis system also has the following advantages:

1. The system can realize the automatic processing of user comments, which greatly improves the efficiency and accuracy of data processing. Compared with traditional manual analysis, the system can extract key information more quickly for sentiment analysis and trend prediction.

W. Wang · J. Niu (✉) · X. Ling · J. Feng
School of Artificial Intelligence, Shanghai Normal University Tianhua College, Shanghai, China
e-mail: njc2807@sthu.edu.cn

X. Jia et al. (eds.), *2nd International Conference on Big Data, IoT, and Cloud Computing*, EAI/Springer Innovations in Communication and Computing,
https://doi.org/10.1007/978-3-031-93570-1_24

2. The system can provide multi-dimensional analysis reports to help restaurant management fully understand the overall evaluation and specific opinions of customers on the restaurant. These reports not only include an evaluation of the overall service quality of the restaurant, but also can analyze the details of specific dishes and waiters, and provide specific guidance for the improvement of the restaurant.
3. The restaurant review analysis system is also scalable and flexible. With the continuous advancement of technology and the change of user needs, the system can be continuously upgraded and optimized to adapt to the changes and development of the market.

In summary, the restaurant review analysis system plays an important role in improving the service quality and optimizing the customer experience in the catering industry. In the future, with the continuous development of technology and the continuous expansion of application scenarios, the system will play a more important role in the catering industry and provide strong support for the sustainable development of restaurants.

2 Methods

2.1 Development Process

The system development process is shown in Fig. 1. It introduces the basic process from system development to front-end display, from data collection to data cleaning, data preprocessing, then sentiment analysis, and finally to front-end display. The following is the detailed process.

The first step is to crawl the data from the website and use the crawler software to crawl the operation. After the crawling operation, the restaurant user comment data that has not been processed is obtained.

The second step is to clean the unprocessed data. First, the text document obtained in the previous step is converted into a text table form, in which the stop thesaurus is used to delete redundant symbols, useless words, and words that affect the machine's judgment in the data.

The third step is to select the appropriate library to call the processing data and use the Jieba library to perform Chinese word segmentation on the remaining data. After processing, a clean text table data suitable for machine processing is obtained. In this process, the vocabulary cloud library can be used to visually display keywords, and the user's most important actual needs such as environment, food quality, and service attitude can be fully understood. The step of preprocessing is an indispensable part of data analysis. It can ensure the accuracy and reliability of the data and improve the efficiency and accuracy of subsequent analysis.

The fourth step uses the SnowNLP library, which can analyze the sentiment of Chinese. After debugging the code, the positivity and scoring of each comment are

Fig. 1 Flow Chart

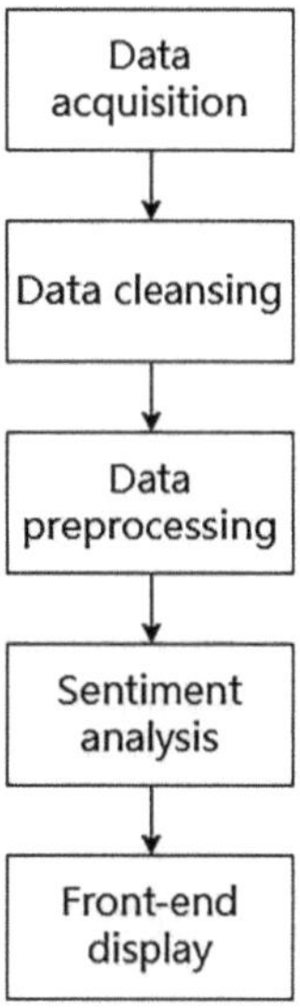

obtained. It is divided from 0 to 1. Judgments close to 1 are negative; judgments close to 0 are positive. Then use the data of all the comments to plot to get a dot matrix that can be well observed. You can use this dot matrix to roughly observe the approximate data distribution. After that, I count all the comments that are well received (data with a positivity greater than 0.5 and less than or equal to 1) to calculate the restaurant's Praise rate. Use this praise rate to evaluate whether the restaurant is worthy of diners, or whether it has been sought after or bad reviews by users.

The fifth step is to use PyQt for front-end display of the project. PyQt is a binding framework for Python and GUI. PyQt provides a variety of page layouts and can directly display text tables, so PyQt is used for front-end display.

2.2 *Data Collection*

With the popularity of the mobile Internet, more and more consumers are accustomed to posting reviews of products or services on social media to express their inner emotions. The evaluation of these spontaneous emotional expressions is more authentic [4]. Therefore, the system uses the Public Comment Network platform to locate target restaurants, filter, and collect their user comment data to ensure legal compliance, and organize and analyze them for subsequent use.

2.3 Data Preprocessing

Due to the different nature of the Chinese text and the English text, there is no space between the words of the text to separate, so the preprocessing of the Chinese text is very important [5].

Chinese word segmentation is a basic step in Chinese text processing and a basic module for Chinese human-computer natural language interaction. When performing Chinese natural language processing, word segmentation is usually required first.

Use the Jieba library to perform Chinese word segmentation on the processed online comment data, so that the data can be processed to obtain a data that contains only Chinese word segmentation.

2.4 Data Cleaning

Generally, there is more dirty data in the online comment data collected by Web crawlers. In order to avoid errors caused by insufficient data specifications, data cleaning operations are required. In order to obtain data suitable for machine reading, Chinese word segmentation is used first. After the word segmentation of the Chinese text content is completed, there will be many words in the comment content that need to be filtered. These words may be redundant information, or they may be meaningless or unrepresentative content. After filtering, it is easy to reduce the storage space and improve the accuracy of retrieving information. These words, such as "of," "and," "you," "yes," and "just," are often used as general terms for evaluating articles, but they do not have much research value. If these words are not filtered out, it is very likely that they will become a new feature, which will affect the running time of the model. Therefore, to eliminate this phenomenon, it is necessary to download an unusable vocabulary dictionary in advance, and then match the vocabulary with the words in the vocabulary dictionary one by one, and eliminate it after matching [6]. The thesaurus of stop words used in this system contains more than 5000 characters and stop words.

The stop thesaurus contains some useless symbols such as symbols and emoticons that affect the machine's reading of data. To filter out these useless symbols and get rid of them, you must first select these words before you can delete them. Otherwise, if these useless and messy information including emoticons is directly handed over to the machine for processing, it will definitely affect the machine's misjudgment and lead to the accuracy of sentiment analysis in the later stage, which will lead to large experimental errors, resulting in the restaurant users' comments cannot be analyzed intuitively. Removing these useless characters not only facilitates the observation of the data, but also makes the accuracy of the experiment higher, killing two birds with one stone.

2.5 Sentiment Analysis

Sentiment analysis has attracted much attention in the past 10 years due to the rapid development of Internet information. How to extract effective user opinions and emotional attitudes from massive amounts of text information has become a meaningful research work. Among them, the extraction of emotional dictionaries and the classification of emotional tendencies are two typical and important research directions [7].

This system mainly uses the functions of the SnowNLP toolkit. SnowNLP has the functions of word segmentation and sentiment analysis. The sentiment analysis of paragraphs of Chinese sentences is carried out to analyze the positivity of the sentences. SnowNLP uses machine learning and natural language processing techniques to perform emotional analysis of text, classify it into positive, negative, or neutral emotional categories, and give corresponding emotional scores. Emotion is a real number between 0 and 1, where 0 represents a strong negative emotion, 1 represents a strong positive emotion, and the middle value indicates a neutral or weaker emotional tendency [8].

Calling SnowNLP(data) creates an instance of the SnowNLP class, with the input data—typically a text string—passed as a parameter for sentiment analysis. By accessing the sentiments attribute, the built-in Bayesian classifier is invoked to perform sentiment analysis, returning a sentiment score between 0 and 1. A value closer to 1 indicates a more positive sentiment, while a value near 0 suggests a negative sentiment [9].

3 Results

3.1 System Function Display

As shown in Fig. 2, the functions of this system are mainly divided into four modules. The functions of each module are as follows:

The main purpose of the word cloud module in this system is to allow users to intuitively observe what high-frequency vocabulary is. They can have a basic understanding of a restaurant. The specific performance is to make the font of high-frequency vocabulary larger and easier to observe through cloud map display; to make low-frequency vocabulary smaller and highlight the difference from high-frequency vocabulary. Let the characteristics of the restaurant show. At the same time, the word cloud map has rich visual effects, which can attract users' attention and stimulate their interest. This visual appeal helps to increase users' attention and participation in text data; allows users to arouse interest in observing text information.

At present, there are two common methods of sentiment polarity analysis: the method based on sentiment dictionary and the method based on machine learning,

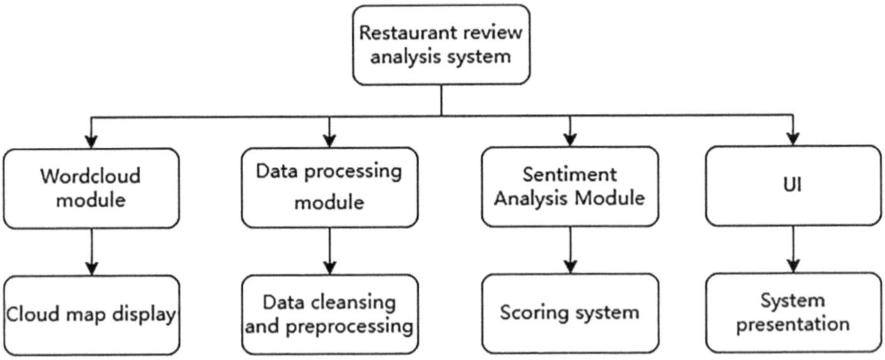

Fig. 2 System function module

this paper uses deep learning technology to conduct sentiment analysis on restaurant reviews, aiming to accurately identify the sentiment tendency in reviews and provide valuable feedback for restaurant operators [10].

In this system, the data processing module is the core component used to process and analyze data in the system. It is usually responsible for receiving the original data, cleaning, converting, integrating, and other operations on the data for subsequent analysis and processing. Regarding data cleaning, the system optimizes text processing by deleting useless words and emoticons in the discontinued thesaurus; then the data is preprocessed and the previously cleaned text files are processed into more convenient text tables.

This system mainly uses the SnowNLP toolkit to perform sentiment analysis on the processed data. The system can use the user's comment information to score. The score is from 0 to 1, and 1 point is a full score. The closer the user's comment score is to 1, the more satisfied the user is with the restaurant. The popularity of the restaurant can also be determined by the user's score.

In the dot matrix drawn by this system, its ordinate represents the user's evaluation score, and its abscissa represents the number of user comments. By observing the distribution of points in the picture, you can observe the degree of good or bad reviews of the restaurant. The distribution of data can be effectively observed to analyze the situation of user comments, and it can be seen that user comments are mixed. All data were counted to calculate the overall proportion of reviews that tended to 1 to obtain the restaurant's praise rate. It was found that the positive data accounted for only 62%, and the restaurant reached the passing line.

Using PyQt to handle the front-end interface, users can click on the name of the restaurant they want to go to observe the comment data left by the user before the restaurant.

Open the selected file and you can clearly observe every data. The result is composed of three columns. The first column represents the serial number, the second column is the specific content of each comment by the user, the third column shows the predicted value of the system, which is the system's score for the sentiment analysis of user comments, and the third column shows the system's score for the user. Classification, comments with not less than 0.5 points will be judged as positive comments; comments with less than 0.5 points will be judged as negative comments.

It is found that the positive data only accounts for 62%, so the restaurant has reached the passing mark.

4 Conclusion

This system effectively bridges the information gap between restaurant owners and customers, enabling owners to gain valuable insights into user preferences and needs. By analyzing customer reviews, restaurant owners can enhance the quality of their services, leading to increased popularity and a broader customer base. Additionally, this platform empowers customers to enjoy improved dining experiences and better service, ultimately fostering a more satisfying interaction between the two parties.

References

1. Hanyang, L., Zhini, L., Xudong, L., et al.: The influencing mechanism of online word-of-mouth: the mediation of trust and the adjustment of gender and involvement. J. Syst. Manag. **3**, 1 (2019)
2. Junping, W., Yue, W., Jing, Y.: The influence of the source characteristics and information characteristics of online word-of-mouth on the purchase intention of health food. Enterprise Econom. **4**, 1 (2019)
3. Wenming, Z., Shaojie, C., Wang, X., et al.: Multi-attribute behavior decision-making model of network consumers based on prospect theory. J. Manag. Eng. **3**, 125 (2019)
4. Shiwei, X., Jiangang, P., Chunmeng, L.: Big data analysis of internet celebrity brands: a case study of "Nai Xue's tea". J Chongqing Technol. Business Univ. (Social Sciences). **39**(6), 47–58 (2022)
5. Zilin, Z.: Analysis of User Satisfaction of Catering Service Based on Dianping Text, p. 15. Zhongnan University of Economics and Law, Hubei (2020)
6. Changcheng, W.: Research on Catering Project Experience Based on Theme Modeling–A Case Study of Haidilao's Review Data. Harbin University of Commerce (2023.23-24)
7. Jia, F.: Sentiment Analysis Technology of Restaurant Review Based on Semi-Supervised Learning. Huazhong University of Science and Technology, Hubei (2016) (in Chinese)
8. Yuqi, X., Ta, Y.: Deep web crawler and data analysis based on reverse technology. Softw. Eng. **26**(12), 41–45 (2023)
9. Jiayi, Y., Xiaoying, L., Shouwu, H.: Research on user sentiment analysis based on Mobile phone review text. Popular Sci Technol. **24**(9), 22–24 (2022)
10. Smitha, R., Venkatesh, P.: Sentiment analysis of restaurant reviews using deep learning. Int. J. Adv. Res. Comput. Sci. Soft. Eng. **10**(5), 12–13 (2020)

Index

X. Jia et al. (eds.), *2nd International Conference on Big Data, IoT, and Cloud Computing*, EAI/Springer Innovations in Communication and Computing,
https://doi.org/10.1007/978-3-031-93570-1

MIX
Papier aus verantwortungsvollen Quellen
Paper from responsible sources
FSC® C105338

If you have any concerns about our products,
you can contact us on
ProductSafety@springernature.com

In case Publisher is established outside the EU,
the EU authorized representative is:
Springer Nature Customer Service Center GmbH
Europaplatz 3, 69115 Heidelberg, Germany

Printed by Libri Plureos GmbH
in Hamburg, Germany